Control Theory of Systems Governed by Partial Differential Equations

Academic Press Rapid Manuscript Production

Control Theory of Systems Governed by Partial Differential Equations

EDITORS:

A.K. AZIZ

University of Maryland
Baltimore County
Baltimore, Maryland

J.W. WINGATE

Naval Surface Weapons Center
White Oak, Silver Spring, Maryland

M.J. BALAS

C. S. Draper Laboratory, Inc.
Cambridge, Massachusetts

ACADEMIC PRESS, INC. 1977

New York San Francisco London

A Subsidiary of Harcourt Brace Jovanovich, Publishers

ACADEMIC PRESS, INC.
111 Fifth Avenue, New York, New York 10003

United Kingdom Edition published by
ACADEMIC PRESS, INC. (LONDON) LTD.
24/28 Oval Road, London NW1

Library of Congress Cataloging in Publication Data

Conference on Control Theory of Systems Governed by Partial Differential Equations, Naval Surface Weapons Center (White Oak), 1976.
Control theory of systems governed by partial differential equations.

Includes bibliographies and index.
1. Control theory–Congresses. 2. Differential equations, Partial–Congresses. I. Aziz, Abdul Kadir. II. Wingate, John Walter. III. Balas, Mark John. IV. Title.
QA402.3.C576 629.8′312′01515353 76-55305
ISBN 0–12–068640–6

PRINTED IN THE UNITED STATES OF AMERICA

Contents

List of Contributors vii

Preface ix

REMARKS ON THE THEORY OF OPTIMAL CONTROL OF DISTRIBUTED SYSTEMS 1
J. L. Lions

STOCHASTIC FILTERING AND CONTROL OF LINEAR SYSTEMS: A GENERAL THEORY 105
A. V. Balakrishnan

DIFFERENTIAL DELAY EQUATIONS AS CANONICAL FORMS FOR CONTROLLED HYPERBOLIC SYSTEMS WITH APPLICATIONS TO SPECTRAL ASSIGNMENT 119
David L. Russell

THE TIME OPTIMAL PROBLEM FOR DISTRIBUTED CONTROL OF SYSTEMS DESCRIBED BY THE WAVE EQUATION 151
H. O. Fattorini

SOME MAX-MIN PROBLEMS ARISING IN OPTIMAL DESIGN STUDIES 177
Earl R. Barnes

VARIATIONAL METHODS FOR THE NUMERICAL SOLUTIONS OF FREE BOUNDARY PROBLEMS AND OPTIMUM DESIGN PROBLEMS 209
O. Pironneau

SOME APPLICATIONS OF STATE ESTIMATION AND CONTROL THEORY TO DISTRIBUTED PARAMETER SYSTEMS 231
W. H. Ray

NUMERICAL SOLUTION OF THE TRANSONIC EQUATION BY THE FINITE ELEMENT METHOD VIA OPTIMAL CONTROL 265
M. O. Bristeau, R. Glowinski, and O. Pironneau

List of Contributors

A. V. BALAKRISHNAN, University of California Los Angeles, California 90024

EARL R. BARNES, IBM Thomas J. Watson Research Center, Yorktown Heights, New York 10598

M. O. BRISTEAU, IRIA/LABORIA, Domaine de Voluceau, 78 Rocquencourt, France

H. O. FATTORINI, Departments of Mathematics and Systems Science, University of California, Los Angeles, California 90024

R. GLOWINSKI, IRIA/LABORIA, Domaine de Voluceau, 78 Rocquencourt, France

J. L. LIONS, IRIA/LABORIA, Domaine de Voluceau, 78 Rocquencourt, France

O. PIRONNEAU, IRIA/LABORIA, Domaine de Voluceau, 78 Rocquencourt, France

W. H. RAY, Department of Chemical Engineering, State University of New York, Buffalo, New York 14214

DAVID L. RUSSELL, Department of Mathematics, University of Wisconsin, Madison, Wisconsin 53706

Preface

These proceedings contain lectures given at the Conference on Control Theory of Systems Governed by Partial Differential Equations held at the Naval Surface Weapons Center (White Oak), Silver Spring, Maryland on May 3-7, 1976.

Most physical systems are intrinsically spatially distributed, and for many systems this distributed nature can be described by partial differential equations. In these distributed parameter systems, control forces are applied in the interior or on the boundary of the controlled region to bring the system to a desired state. In systems where the spatial energy distribution is sufficiently concentrated, it is sometimes possible to approximate the actual distributed system by a lumped parameter (ordinary differential equation) model. However, in many physical systems, the energy distributions are widely dispersed and it is impossible to gain insight into the system behavior without dealing directly with the partial differential equation description.

The purpose of this conference was to examine the control theory of partial differential equations and its application. The main focus of the conference was provided by Professor Lions' tutorial lecture series—Theory of Optimal Control of Distributed Systems—with the many manifestations of the theory and its applications appearing in the presentations of the other invited speakers: Professors Russell, Pironneau, Barnes, Fattorini, Ray, and Balakrishnan.

We wish to thank the invited speakers for their excellent lectures and written summaries. All who were present expressed their satisfaction with the range and depth of the topics covered.

There was strong interaction among the participants, and we hope these published proceedings reflect some of the coherence achieved. We appreciate the contributions of all the attendees and the patience shown with any fault of organization of which we may have been guilty.

We thank the Office of Naval Research for their financial support of this conference. Finally, special thanks are due Mrs. Nancy King on whom the burden of typing this manuscript fell.

"REMARKS ON THE THEORY OF OPTIMAL CONTROL OF DISTRIBUTED SYSTEMS"

J. L. Lions

Introduction

These notes correspond to a set of lectures given at the Naval Surface Weapons Center, White Oak Laboratory, White Oak, Maryland 20910, May 3 through May 7, 1976.

In these notes we present a partial survey of some of the trends and problems in the theory of optimal control of distributed systems.

In Chapter 1 we present some more or less standard material, to fix notations and ideas; some of the examples presented there can be thought of as simple exercises.

In Chapter 2 we recall some known facts about duality methods, together with the connection between duality, regularization and penalty (we show this in an example); we also give in this chapter a recent result of H. Brezis and I. Ekeland (actually a particular use of it) giving a variational principle for, say, the heat equation (a seemingly long standing open question, which admits a very simple answer).

Chapter 3 gives an introduction to some asymptotic methods which can be useful in control theory; we give an example of the connection between "cheap control" and singular perturbations; we show next how the "homogeneization" procedure, in composite materials, can be used in optimal control.

In Chapter 4 we study the systems which are non-linear or whose state equation is an eigenvalue or an eigenfunction; we present two examples of this situation; we consider then an example where the control variable is a function which appears in the coefficients of the highest derivatives and next we consider an example where these two properties (control in the highest derivatives and state = eigenfunction) arise simultaneously. We study then briefly the control of free surfaces and problems where the control variable is a geometrical argument (such as in optimum design). We end this chapter with several open questions.

In Chapter 5 we give a rather concise presentation of the use of mixed finite elements for the numerical computation of optimal controls. For further details we refer to Bercovier [1].

All the examples presented here are related to, or motivated by, specific applications, some of them being referred to in the Bibliography.

We do not cover here, among other things:

the controllability problems (cf. Fattorini [1], Russell [1] in these proceedings), the stability questions, such as Feedback Stabilization (let us mention in this respect Kwan and K. N. Wang [1], J. Sung and C. Y. Yii [1], and Sakawa and Matsushita [1]; cf. also Saint Jean Paulin [1]);

the identification problems for distributed systems, which can be put in the framework of optimal control theory, and for which we refer to G. Chavent [1], G. Chavent and P. Lemonnier [1] (for applications to geological problems), to G. I. Marchuk [1] (for applications in meteorology and oceanography), to Begis and Crepon [1] (for applications to oceanography), to J. Blum (for applications to plasma physics); cf. also the surveys Polis and Goodson [1] and Lions [11];

problems with delays, for which we refer to Delfour and Mitter [1] and to the bibliography therein;

multicriteria problems, and stochastic problems.

For other applications than those indicated here, let us refer to the recent books Butkovsky [1], Lurie [1], Ray and Lainiotis [1], P. K. C. Wang [1].

The detailed plan is as follows:

Chapter 1. Optimality conditions for linear-quadratic systems.

1. A model example.

   1.1 Orientation

1.2 The state equation
1.3 The cost function. The optimal control problem
1.4 Standard results
1.5 Particular cases
2. A noninvertible state operator.
2.1 Statement of the problem
2.2 The optimality system
2.3 Particular cases
2.4 Another example
2.5 An example of "parabolic-elliptic" nature
3. An evolution problem
3.1 Setting of the problem
3.2 Optimality system
3.3 The "no constraints" case
3.4 The case when $U_{ad} = \{v | v \geq 0$ a.e. on $\Sigma\}$.
3.5 Various remarks
4. A remark on sensitivity reduction
4.1 Setting of the problem
4.2 The optimality system
5. Non well set problems as control problems
5.1 Orientation
5.2 Formulation as a control problem
5.3 Regularization method
Chapter 2. Duality methods.
1. General considerations
1.1 Setting of the problem
1.2 A formal computation
2. A problem with constraints on the state
2.1 Orientation
2.2 Setting of the problem
2.3 Transformation by duality
2.4 Regularized dual problem and generalized problem
3. Variational principle for the heat equation
3.1 Direct method
3.2 Use of duality

Chapter 3. Asymptotic methods.

1. Orientation
2. Cheap control. An example
   2.1 Setting of the problem
   2.2 A convergence theorem
   2.3 Connection with singular perturbations
3. Homogeneization
   3.1 A model problem
   3.2 The homogeneized operator
   3.3 A convergence theorem

Chapter 4. Systems which are not of the linear quadratic type.

1. State given by eigenvalues or eigenfunctions
   1.1 Setting of the problem
   1.2 Optimality conditions
   1.3 An example
2. Another example of a system whose state is given by eigenvalues or eigenfunctions
   2.1 Orientation
   2.2 Statement of the problem
   2.3 Optimality conditions
3. Control in the coefficients
   3.1 General remarks
   3.2 An example
4. A problem where the state is given by an eigenvalue with control in the highest order coefficients
   4.1 Setting of the problem
   4.2 Optimality conditions
5. Control of free surfaces
   5.1 Variational inequalities and free surfaces
   5.2 Optimal control of variational inequalities
   5.3 Open questions
6. Geometrical control variables
   6.1 General remarks
   6.2 Open questions

Chapter 5. Remarks on the numerical approximation of problems of optimal control

1. General Remarks
2. Mixed finite elements and optimal control
   - 2.1 Mixed variational problems
   - 2.2 Regularization of mixed variational problems
   - 2.3 Optimal control of mixed variational systems
   - 2.4 Approximation of the optimal control of mixed variational systems

Chapter 1

Optimality Conditions for Linear-Quadratic Systems

1. A Model Example

1.1 Orientation

We give here a very simple example, which allows us to introduce a number of notations we shall use in all that follows.

1.2 The state equation

Let Ω be a bounded open set in R^n, with smooth boundary Γ.

Let A be a second order elliptic operator, given by

$$A\phi = -\sum_{i,j=1}^{n} \frac{\partial}{\partial x_i}\left(a_{ij}(x)\frac{\partial\phi}{\partial x_j}\right) + \sum_{j=1}^{n} a_j(x)\frac{\partial\phi}{\partial x_j} + a_0(x)\phi, \tag{1.1}$$

where the functions a_{ij}, a_j, a_0 belong to $L^\infty(\Omega)$; we introduce the Sobolev space $H^1(\Omega)$:

$$H^1(\Omega) = \left\{\phi \mid \phi, \frac{\partial\phi}{\partial x_i} \in L^2(\Omega)\right\}, \tag{1.2}$$

provided with the norm

$$\|\phi\| = \left(|\phi|^2 + \sum \left|\frac{\partial\phi}{\partial x_i}\right|^2\right)^{1/2}, \tag{1.3}$$

where

$$|\phi| = \left(\int_\Omega \phi^2 dx\right)^{1/2} = \text{norm in } L^2(\Omega), \tag{1.4}$$

(all functions are assumed to be real valued); provided with (1.3), $H^1(\Omega)$ is a Hilbert space; for $\phi, \psi \in H^1(\Omega)$ we set

$$a(\phi,\psi) = \sum\int_\Omega a_{ij}\frac{\partial\phi}{\partial x_j}\frac{\partial\psi}{\partial x_i}dx + \sum\int_\Omega a_j\frac{\partial\phi}{\partial x_j}\psi\, dx + \int_\Omega a_0\phi\psi\, dx. \tag{1.5}$$

We assume A to be $H^1(\Omega)$ - elliptic, i.e.

$$a(\phi,\phi) \geq \alpha\|\phi\|^2, \quad \alpha > 0, \quad \forall\phi \in H^1(\Omega). \tag{1.6}$$

The state equation in its variational form is now:

(1.7) $$a(y,\psi) = (f,\psi) + \int_\Gamma v\psi d\Gamma \quad \forall\psi\in H^1(\Omega) ,$$

where $(f,\psi) = \int_\Omega f\psi dx$, f given in $L^2(\Omega)$, and where in (1.7) the "control variable" v is given in $L^2(\Gamma)$.

We recall that $\forall\psi\in H^1(\Omega)$ one can uniquely define the "trace" of ψ on Γ ; it is an element of $L^2(\Gamma)$ (actually of a smaller space $H^{1/2}(\Gamma)$) and the mapping

$$\psi \longrightarrow \psi|\Gamma$$

is continuous from $H^1(\Omega) \longrightarrow L^2(\Gamma)$.

Therefore the right hand side in (1.7) defines a continuous linear form on $H^1(\Omega)$, so that, by virtue of (1.6):

(1.8) $$\begin{cases} \text{Equation (1.7) admits a unique solution, denoted} \\ \text{by } y(v);\ y(v) \in H^1(\Omega) \text{ and the mapping} \\ v \longrightarrow y(v) \text{ is affine continuous from } L^2(\Gamma) \longrightarrow H^1(\Omega) . \end{cases}$$

The interpretation of (1.7) is as follows:

(1.9) $$Ay(v) = f \text{ in } \Omega$$

(1.10) $$\frac{\partial y(v)}{\partial\nu_A} = v \text{ on } \Gamma ,$$

where $\frac{\partial\phi}{\partial\nu_A} = \sum a_{ij}(x) \frac{\partial\phi}{\partial x_j} \cos(\nu,x_i)$, ν = unit normal to Γ directed toward the exterior of Ω ; of course, under only the hypothesis that $a_{ij} \in L^\infty(\Omega)$, (1.10) is formal; in case $a_{ij} \in W^{1,\infty}(\Omega)$ (i.e. $\frac{\partial a_{ij}}{\partial x_k} \in L^\infty(\Omega)\ \forall\ k$) , then one can show that $y(v)\in H^2(\Omega)$ ([1]) and

---

([1]) $$H^2(\Omega) = \left\{ \phi \mid \phi, \frac{\partial\phi}{\partial x_i}, \frac{\partial^2\phi}{\partial x_i\, \partial x_j} \in L^2(\Omega) \right\}.$$

(1.10) becomes precise. In the general case one says that $y = y(v)$, solution of (1.7) is a weak solution of (1.9)(1.10) . #

We shall call (1.7) (or (1.9)(1.10)) the state equation, $y(v)$ being the state of the system.

1.3 The cost function. The optimal control problem.

To each control v we associate a cost $J(v)$ defined by

$$J(v) = \int_\Gamma |y(v)-z_d|^2 d\Gamma + N \int_\Gamma v^2 d\Gamma , \tag{1.11}$$

where z_d is given in $L^2(\Gamma)$ and where N is a given positive number.

Let v belong to a subset U_{ad} of $L^2(\Gamma)$ (the set of admissable controls); we assume

(1.12) U_{ad} is a closed non-empty convex subset of $L^2(\Gamma)$.

We shall refer to the case $U_{ad} = L^2(\Gamma)$ as the "no constraint" case.

The problem of optimal control is now

$$\text{find inf } J(v) , \quad v \in U_{ad} . \tag{1.13}$$

1.4 Standard results. (cf. Lions [1])

Problem (1.13) admits a unique solution u (the optimal control).

This optimal control u is characterized by

$$\begin{cases} (J'(u), v-u) \geq 0 \quad \forall v \in U_{ad} , \\ u \in U_{ad} \end{cases} \tag{1.14}$$

where $(J'(u), v) = \frac{d}{d\xi} J (u+\xi v)\Big|_{\xi=0}$ (this derivative exists).

The condition (1.14) which gives the necessary and sufficient condition for u to minimize J over U_{ad} is (a particular case of a Variational Inequality (V.I.).

An explicit (and trivial) computation of $J'(u)$ gives (after dividing by 2)

$$(1.15)\quad \begin{cases} \int_\Gamma (y(u)-z_d)\ (y(v)-y(u))\ d\Gamma + N \int_\Gamma u(v-u)\ d\Gamma \geq 0 \\ \forall v \in U_{ad},\ u \in U_{ad}. \quad \# \end{cases}$$

<u>Transformation of</u> (1.15) <u>by using the adjoint state</u>.

In order to transform (1.15) in a more convenient form, we introduce the <u>adjoint state</u> p defined by

$$(1.16)\quad \begin{cases} A^* p = 0 \quad \text{in} \quad \Omega , \\ \dfrac{\partial p}{\partial \nu_{A^*}} = y - z_d \quad \text{on} \quad \Gamma \end{cases}$$

where we set $y(u) = y$, A^* = adjoint of A .

The <u>variational form</u> of (1.16) is

$$(1.17)\quad a^* (p,\psi) = \int_\Gamma (y-z_d)\psi\ d\Gamma \ \ \forall \psi \in H^1(\Omega)$$

where we define

$$(1.18)\quad a^* (\phi,\psi) = a(\psi,\phi) .$$

Let us set

$$X = \int_\Gamma (y-z_d)\ (y(v)-y) d\Gamma \ ;$$

by taking $\psi = y(v)-y$ in (1.17) we obtain

$$X = a^* (p,y(v)-y) = a(y(v)-y,p) = (\text{by using (1.7)}) = \int_\Gamma (v-u)p\ d\Gamma$$

and (1.15) becomes

$$(1.19)\quad \begin{cases} \int_\Gamma (p+Nu)\ (v-u)\ d\Gamma \geq 0 \qquad v \in U_{ad}, \\ u \in U_{ad}. \qquad \# \end{cases}$$

We can summarize as follows the results obtained so far: the optimal control u of (1.13) is characterized through the unique solution $\{y, p, u\}$ of the optimality system given by:

$$\text{(1.20)} \quad \begin{cases} Ay = f \quad \text{in} \quad \Omega , \\ A^*p = 0 \quad \text{in} \quad \Omega , \\ \dfrac{\partial y}{\partial \nu_A} = u \ , \ \dfrac{\partial p}{\partial \nu_{A^*}} = y - z_d \quad \text{on} \quad \Gamma , \\ \int_\Gamma (p+Nu)(v-u)d\Gamma \geq 0 \ \forall v \in U_{ad}, \ u \in U_{ad} . \end{cases}$$

1.5 Particular cases.

1.5.1 The case without constraints.

If $U_{ad} = L^2(\Gamma)$, the last condition in (1.20) reduces to

$$\text{(1.21)} \quad p + Nu = 0 .$$

Then one solves the system of elliptic equations:

$$\text{(1.22)} \quad \begin{cases} Ay = f, \ A^*p = 0 \quad \text{in} \quad \Omega , \\ \dfrac{\partial y}{\partial \nu_A} + \dfrac{1}{N} p = 0, \ \dfrac{\partial p}{\partial \nu_{A^*}} = y - z_d \quad \text{on} \end{cases}$$

and u is given by (1.21).

1.5.2 $U_{ad} = \{v | \ v \geq 0 \text{ a.e. on } \Gamma\}$.

In case U_{ad} is given by 1.5.2, the last condition (1.20) is equivalent to

$$\text{(1.23)} \quad u \geq 0, \ p + Nu \geq 0, \ u(p+Nu) = 0$$

i.e.

$$\text{(1.24)} \quad u = \sup (0, - \frac{p}{N}) = \frac{1}{N} p^- .$$

Then one solves the system of non-linear elliptic equations:

$$(1.25) \quad \begin{cases} Ay = f, \ A^*p = 0 \quad \text{in} \quad \Omega , \\ \dfrac{\partial y}{\partial \nu} - \dfrac{1}{N} p^- = 0, \quad \dfrac{\partial p}{\partial \nu_{A^*}} = y - z_d \quad \text{on} \quad \Gamma \end{cases}$$

and u is given by (1.24).

Remark 1.1

By virtue of the way we found (1.25), this system admits a unique solution $\{y,p\}$.

Remark 1.2

We have two parts on Γ :

$$\Gamma^- = \{x \mid x \varepsilon \Gamma, \ p(x) \leq 0\}, \ \Gamma^+ = \{x \mid x \varepsilon \Gamma, \ p(x) > 0\}$$

(these regions are defined up to a set of measure 0 on Γ) and $u = 0$ on Γ^+ . The interface between Γ^- and Γ^+ can be thought of as a free surface or as a commutation line. #

Remark 1.3

For interesting examples related to the above techniques, we refer to Boujot, Morera and Temam [1]. #

2. A non invertible state operator.

2.1 Statement of the problem

In order to simplify the exposition we shall assume that

$$(2.1) \qquad A = - \Delta$$

but what we are going to say readily extends to the case when A is any self-adjoint elliptic operator of any order (or to a self-adjoint system).

We suppose that the state y is given by

$$(2.2) \quad \begin{cases} -\Delta y = f - v \quad \text{in} \quad \Omega , \\ \dfrac{\partial y}{\partial \nu} = 0 \quad \text{on} \quad \Gamma . \end{cases}$$

But now if A denotes the unbounded operator $-\Delta$ with domain $\{\phi | \phi \varepsilon H^1(\Omega), \Delta\phi \varepsilon L^2(\Omega), \frac{\partial\phi}{\partial\nu} = 0 \text{ on } \Gamma\}$, 0ε spectrum of A so that A is not invertible; but a necessary and sufficient condition for (2.2) to admit a solution is

$$(2.3) \qquad (f-v,1) = 0$$

and then (2.2) admits an infinite number of solutions; we uniquely define $y(v)$ by adding, for instance, the condition

$$(2.4) \qquad M(y(v)) = 0 \ ,$$

where $M(\phi) = \frac{1}{|\Omega|} \int_\Omega dx$, $|\Omega|$ = measure of Ω .

Summing up: we consider control functions v which satisfy (2.3); then the state $y(v)$ of the system is given as the solution of (2.2) (2.4). #

The cost function is given by

$$(2.5) \qquad J(v) = \int_\Gamma |y(v)-z_d|^2 \, d\Gamma + N \int_\Omega v^2 dx \ .$$

We consider

$$(2.6) \qquad \begin{cases} U_{ad} = \text{closed convex subset of } L^2(\Gamma) \text{ and of the (linear)} \\ \text{set defined by (2.3)} \ , \end{cases}$$

and we want again to solve

$$(2.7) \qquad \inf J(v), \ v \varepsilon U_{ad} \ .$$

2.2 The optimality system.

One easily checks that problem (2.7) admits a unique solution u, which is characterized by (we set $y(u) = y$):

$$(2.8) \qquad \begin{cases} \int_\Gamma (y-z_d)(y(v)-y)d\Gamma + N(u,v-u) \geq 0 \quad \forall v \varepsilon U_{ad}, \\ u \varepsilon U_{ad} \ . \qquad \# \end{cases}$$

We introduce now the adjoint state p as the solution of

$$(2.9)\qquad \begin{cases} -\Delta p = \frac{-1}{|\Omega|} \int_\Gamma (y-z_d)d\Gamma \quad \text{in } \Omega , \\ \frac{\partial p}{\partial \nu} = y-z_d \quad \text{on } \Gamma , \\ M(p) = 0 . \end{cases}$$

We remark that (2.9) admits a unique solution.

If we take the scalar product of the first equation in (2.9) with $y(v) - y$, we obtain

$$(-\Delta p, y(v)-y) = \frac{-1}{|\Omega|} \int_\Gamma (y-z_d)d\Gamma \ (1, y(v)-y) = 0$$

$$\text{(by virtue of (2.4))} = - \int_\Gamma \frac{\partial p}{\partial \nu} (y(v)-y)d\Gamma +$$

$$+ (p, -\Delta(y(v)-y)) = - \int_\Gamma (y-z_d)(y(v)-y)d\Gamma + (p,-(v-u))$$

(the use we make here of Green's formula is justified; one has just to think of the variational formulation of these equations).
Then (2.8) reduces to

$$(2.10)\qquad (-p+Nu,v-u) \geq 0 \quad \forall v\in U_{ad}, \ u\in U_{ad} .$$

Summarizing, we have: the optimal control u , unique solution of (2.7), is characterized by the solution $\{y, p, u\}$ of the optimality system:

$$\begin{cases} -\Delta y = f-u, \ -\Delta p = \frac{-1}{|\Omega|} \int_\Gamma (y-z_d)d\Gamma \quad \text{in } \Omega , \\ \frac{\partial y}{\partial \nu} = 0, \ \frac{\partial p}{\partial \nu} = y-z_d \quad \text{on } \Gamma , \\ M(y) = 0, \ M(p) = 0, \\ (-p+Nu, v-u) \geq 0 \quad \forall v\in U_{ad} , \ u\in U_{ad} . \end{cases}$$

2.3 Particular cases

Let us suppose that

(2.12) $$U_{ad} = \{v \mid (v,1) = (f,1)\}$$

i.e. the biggest possible choice of U_{ad}.
Then (2.10) is equivalent to

(2.13) $$\begin{cases} -p + Nu = c = \text{constant} \\ \text{and} \quad u \in U_{ad} , \end{cases}$$

i.e.

$$-(p,1) + N(u,1) = c\ |\Omega| = N(f,1) \quad \text{i.e.}$$

(2.14) $$c = N\, M(f) .$$

Then one solves first the system:

(2.15) $$\begin{cases} -\Delta y + \frac{p}{N} = f - Mf , \\ -\Delta p = \frac{-1}{|\Omega|} \int_\Gamma (y - z_d) d\Gamma , \\ \frac{\partial y}{\partial \nu} = 0, \ \frac{\partial p}{\partial \nu} = y - z_d \quad \text{on } \Gamma , \\ M(y) = M(p) = 0 , \end{cases}$$

and then

(2.16) $$u = Mf + \frac{p}{N} . \qquad \#$$

Let us now suppose that

(2.17) $$U_{ad} = \{v \mid v \geq 0 \text{ a.e. in } \Omega, \ (v,1) = (f,1)\} ,$$

under hypothesis

(2.18) $$Mf > 0$$

which implies that U_{ad} does not reduce to $\{0\}$ (case $Mf = 0$) or is not empty (case $Mf < 0$) .

<u>Then the solution of</u> (2.10) <u>is given by</u>

$$u = \frac{p}{N} + Mf + r - Mr \, , \tag{2.19}$$

where

$$\begin{cases} r = (\frac{p}{N} + Mf - \lambda)^- \, , \\ \lambda \varepsilon R, \ \lambda \text{ being a solution of} \\ \lambda = M(r) \, . \end{cases} \tag{2.20}$$

Indeed, let us check first that $\lambda = M(r)$ admits a solution (actually unique if $\int_\Gamma (y-z_d)d\Gamma \neq 0$) , at least assuming that

$$\chi = \frac{p}{N} + Mf \ \varepsilon L^\infty(\Omega) \ ;$$

if we set $\rho(\lambda) = M((\chi-\lambda)^-)$ then $\rho(\lambda)$ is an increasing function, $\rho(\lambda) = 0$ for $\lambda \leq -c$, $\rho(\lambda) = \lambda - M(\chi) = \lambda - Mf$ for λ large enough, hence the result follows; let us notice that u defined by (2.19) does not depend on the choice of λ satisfying (2.20); let us now check that u satisfies (2.10). We can write

$$u = \frac{p}{N} + Mf - \lambda + r = (\frac{p}{N} + Mf - \lambda)^+ \geq 0 \, . \tag{2.21}$$

We have $M(u) = M(f)$, and

$$(-\frac{p}{N} + u, \ v-u) = (Mf - \lambda + r, \ v-u) = (r,v-u) \ ;$$

but $(r, u) = 0$ hence

$$(-\frac{p}{N} + u, \ v-u) = (r,v) \geq 0 \quad \text{since } r \text{ and } v \text{ are} \geq 0 \, ,$$

hence the result follows.

The optimality system is given by

$$(2.22)\quad \begin{cases} -\Delta y = f - (\frac{p}{N} + Mf - \lambda)^{+} , \\ -\Delta p = \frac{-1}{|\Omega|} \int_\Gamma (y-z_d)d\Gamma , \\ \frac{\partial y}{\partial \nu} = 0, \ \frac{\partial p}{\partial \nu} = y-z_d \ \text{ on } \Gamma , \ M(y) = M(p) = 0 , \\ \lambda = M\left((\frac{p}{N} + Mf - \lambda)^{-} \right) . \qquad \# \end{cases}$$

Remark 2.1 Regularity of the optimal control.

It follows from (2.20) or (2.21) that

$$(2.23)\quad u \in H^1(\Omega) ,$$

since $\frac{p}{N} + Mf - \lambda \in H^1(\Omega)$.

Let us also remark that if $z_d \in H^{1/2}(\Gamma)$ then $p \in H^2(\Omega)$ but this does not improve (2.23) . #

Remark 2.2

One can find (2.19) (2.20) by a duality argument. (cf. Chapter 2 for the duality method).

2.4 Another example.

As an exercise, let us consider the state equation

$$(2.24)\quad \begin{cases} -\Delta y = f \ \text{ in } \Omega , \\ \frac{\partial y}{\partial \nu} = v \ \text{ on } \Gamma \end{cases}$$

which admits a set of solutions $\{y + \text{constant}\}$ iff

$$(2.25)\quad - \int_\Gamma v \, d\Gamma = \int_\Omega f \, dx .$$

We define the state $y(v)$ as the solution of (2.24) which satisfies

$$(2.26)\quad M(y) = 0 .$$

If we consider the cost function

$$(2.27) \qquad J(v) = \int_\Gamma |y(v)-z_d|^2 \, d\Gamma + N \int_\Gamma v^2 \, d\Gamma \ ,$$

then the optimality system is given by

$$(2.28) \qquad \begin{cases} -\Delta y = f, \ -\Delta p = \frac{-1}{|\Omega|} \int_\Gamma (y-z_d) d\Gamma \quad \text{in} \quad \Omega \ , \\ \frac{\partial y}{\partial \nu} = u, \ \frac{\partial p}{\partial \nu} = y-z_d \quad \text{on} \quad \Gamma \ , \\ M(y) = M(p) = 0 \ , \\ \int_\Gamma (p+Nu)(v-u) d\Gamma \geq 0 \quad \forall v \in U_{ad}, \ u \in U_{ad} \ , \end{cases}$$

where U_{ad} is a (non-empty) closed convex subset of the set of v's in $L^2(\Omega)$ which satisfy (2.25).

2.5 An example of "parabolic-elliptic" nature

Let us consider now an evolution equation

$$(2.29) \qquad \frac{\partial y}{\partial t} - \Delta y = f-v \quad \text{in} \quad Q = \Omega \times]0, T[\ ,$$

$$f, \ v \in L^2(Q) \ ,$$

with boundary condition

$$(2.30) \qquad \frac{\partial y}{\partial \nu} = 0 \quad \text{on} \quad \Sigma = \Gamma \times]0, T[\ ,$$

and

$$(2.31) \qquad y(0) = y(T) \quad \text{on} \quad \Omega$$

(where y(t) denotes the function $x \to y(x,t)$) .

The equations (2.29) (2.30) (2.31) admit a solution (and actually a set of solutions y + constant) iff

$$(2.32) \qquad \int_Q v \, dx \, dt = \int_Q f \, dx \, dt \ .$$

Let us then define the state of the system as the solution $y(v)$ of (2.29) (2.30) (2.31) such that

$$\int_Q y(v)\, dx\, dt = 0 . \tag{2.33}$$

If the cost function is given by

$$J(v) = \int_Q |y(v)-z_d|^2\, dx\, dt + N \int_Q v^2\, dx\, dt,\ N > 0,\ z_d \in L^2(Q) , \tag{2.34}$$

and if U_{ad} is a (non empty) closed convex subset of the v's in $L^2(Q)$ such that (2.32) holds true, <u>the optimality system is given by</u>

$$\left\{\begin{array}{l} \frac{\partial y}{\partial t} - \Delta y = f - u , \\ -\frac{\partial p}{\partial t} - \Delta p = y-z_d - \frac{1}{|Q|}\int_Q (y-z_d)\, dx\, dt , \\ \frac{\partial y}{\partial \nu} = 0,\ \frac{\partial p}{\partial \nu} = 0 \text{ on } \Sigma , \\ y(0) = y(T),\ p(0) = p(T) , \\ \int_Q y\, dx\, dt = \int_Q p\, dx\, dt = 0 , \\ \int_Q (-p+Nu)(v-u)\, dx\, dt \geq 0 \quad \forall v \in U_{ad} , \\ u \in U_{ad} . \end{array}\right. \tag{2.35}$$

3. <u>An evolution problem</u>.

3.1 <u>Setting of the problem</u>.

We consider now an operator A as in Section 1 (cf. (1.1)); we use the notation (1.5) and we shall assume

(3.1) there exists λ and $\alpha>0$ such that

$$a(\phi,\phi) + \lambda|\phi|^2 \geq \alpha\|\phi\|^2 \quad \forall \phi \in H^1(\Omega)$$

(this condition is satisfied if a_0, $a_j \in L^\infty(\Omega)$ and $a_{ij} \in L^\infty(\Omega)$ such that

$$\sum_{i,j=1}^{n} a_{ij}(x)\,\zeta_i\,\zeta_j \geq \alpha_1 \Sigma\, \zeta_i^2\ ,\ \alpha_1 > 0)\ .$$

We consider the state equation:

(3.2) $$\frac{\partial y}{\partial t} + Ay = f \quad \text{in} \quad Q = \Omega \times]0,T],\ f \varepsilon L^2(Q)\ ,$$

(3.3) $$\frac{\partial y}{\partial \nu} = v \quad \text{on} \quad \Sigma\ ,\ v \varepsilon L^2(\Sigma),\ (^1)$$

(3.4) $$y(0) = y_0 \quad \text{on} \quad \Omega,\ y_0 \varepsilon L^2(\Omega)\ .$$

This problem admits a unique solution which satisfies

(3.5) $$y\ \varepsilon L^2(0,T;H^1(\Omega))\ .$$

(cf. Lions [1] [2] for instance, or Lions-Magenes [1]) #

The variational formulation of this problem is

(3.6) $$(\frac{\partial y}{\partial t}, \psi) + a(y,\psi) = (f,\psi) + \int_\Gamma v\psi d\Gamma \quad \forall \psi \varepsilon H^1(\Omega)$$

with the initial condition (3.4). #

Let the cost function $J(v)$ be given by

(3.7) $$J(v) = \int_\Sigma |y(v)-z_d|^2\, d\Sigma + N \int_\Sigma v^2 d\Sigma,\ z_d \varepsilon L^2(\Sigma),\ N > 0\ ,$$

and let U_{ad} be a (non empty) colsed convex subset of $L^2(\Sigma)$. We consider the problem of minimization:

(3.8) $$\inf J(v),\ v \varepsilon\ U_{ad}\ .$$

3.2 Optimality system.

Problem (3.8) admits a unique solution, say u, which is characterized by

---

(1) We write $\frac{\partial}{\partial \nu}$ instead of $\frac{\partial}{\partial \nu_A}$.

(3.9) $(J'(u), v-u) \geq 0 \ \forall v \in U_{ad}, \ u \in U_{ad}$,

i.e. (where we set $y(u) = y$):

$$\text{(3.10)} \quad \begin{cases} \int_\Sigma (y-z_d)\,(y(v)-y)d\Sigma + N \int_\Sigma u(v-u)d\Sigma \geq 0 \\ \forall v \in U_{ad}, \ u \in U_{ad} \ . \end{cases}$$

In order to simplify (3.10) we introduce as in previous sections the <u>adjoint state</u> p given by

$$\text{(3.11)} \quad \begin{cases} -\frac{\partial p}{\partial t} + A^* p = 0 \quad \text{in} \quad \Omega \ , \\ \frac{\partial p}{\partial \nu^*} = y - z_d \quad \text{on} \quad \Sigma \ , \quad (^1) \\ p(T) = 0 \quad \text{on} \quad \Omega \ . \end{cases}$$

Then

$$\int_\Sigma (y-z_d)\,(y(v)-y)d\Sigma = \int_\Sigma p(v-u)d\Sigma$$

so that (3.10) becomes

(3.12) $\int_\Sigma (p+Nu)\,(v-u)d\Sigma \geq 0 \quad \forall v \in U_{ad}, \ u \in U_{ad}$.

The <u>optimality system</u> is given by

$$\text{(3.13)} \quad \begin{cases} \frac{\partial y}{\partial t} + Ay = f, \ -\frac{\partial p}{\partial t} + A^* p = 0 \quad \text{in} \quad Q \ , \\ \frac{\partial y}{\partial \nu} = u, \ \frac{\partial p}{\partial \nu} = y - z_d \quad \text{on} \quad \Sigma \ , \\ y(0) = y_0, \ p(T) = 0 \quad \text{on} \quad \Omega \ , \\ \int_\Sigma (p+Nu)\,(v-u)d\Sigma \geq 0 \quad \forall v \in U_{ad}, \ u \in U_{ad} \ . \qquad \# \end{cases}$$

---

(1) We write $\frac{\partial}{\partial \nu^*}$ instead of $\frac{\partial}{\partial \nu_{A^*}}$.

3.3 The "no constraints" case.

If we suppose that

$$U_{ad} = L^2(\Sigma) \tag{3.14}$$

then (3.12) reduces to

$$p + Nu = 0 . \tag{3.15}$$

Then one solves first the system in {y, p} :

$$\begin{cases} \frac{\partial y}{\partial t} + Ay = f, \ -\frac{\partial p}{\partial t} + A^* p = 0 \quad \text{in} \quad Q , \\ \frac{\partial y}{\partial \nu} + \frac{1}{N} p = 0, \ \frac{\partial p}{\partial \nu^*} = y - z_d \quad \text{on} \quad \Sigma , \\ y(0) = y_0, \ p(T) = 0 \end{cases} \tag{3.16}$$

and then u is given by (3.15).

Remark 3.1

We obtain a regularity result for $u = -\frac{1}{N} p$; u is an element of $L^2(0,T; H^{1/2}(\Gamma))$ (and one has more, since $p \in L^2(0,T; H^2(\Omega))$ and $\frac{\partial p}{\partial t} \in L^2(Q)$, if we assume more on z_d). #

3.4 The case when $U_{ad} = \{v | v \geq 0 \text{ a.e. on } \Sigma\}$.

In the case when

$$U_{ad} = \{v | \ v \in L^2(\Sigma), \ v \geq 0 \text{ a.e. on } \Sigma\} , \tag{3.17}$$

then (3.12) is equivalent to

$$u \geq 0, \ p + Nu \geq 0, \ u(p+Nu) = 0 \quad \text{on } \Sigma \tag{3.18}$$

i.e.

$$u = \frac{1}{N} p^- . \tag{3.19}$$

Then the optimality system can be solved by solving first the non-linear system in {y, p} given by

$$(3.20)\quad \begin{cases} \frac{\partial y}{\partial t} + Ay = f,\ -\frac{\partial p}{\partial t} + A^* p = 0 \text{ in } Q , \\ \frac{\partial y}{\partial \nu} - \frac{1}{N} p^- = 0,\ \frac{\partial p}{\partial \nu^*} = y - z_d \text{ on } \Sigma , \\ y(0) = y_0,\ p(T) = 0 \text{ on } \Omega , \end{cases}$$

and by using next (3.19).

Remark 3.2

We obtain (as in Remark 3.1) the regularity result on the optimal control:

$$(3.21)\quad u \in L^2(0,T;H^{1/2}(\Gamma)) . \qquad \#$$

3.5 Various remarks.

Remark 3.3

For the "decoupling" of (3.16) and "reduction" of the "two point" boundary value problem in time (3.16) to Cauchy problems for non linear equations (of the Riccati-integro-differential type) we refer to Lions [1] [3] and to recent works of Casti and Ljung [1], Casti [1] Baras and Lainiotis [1](where one will find other references) for the decomposition of the Riccati equation. We refer also to Yebra [1], Curtain and Pritchard [1], Tartar [1].

Remark 3.6

We also refer to Lions, loc. cit, for similar problems for higher order operators A, or operators A with coefficients depending on x and on t ; also for operators of hyperbolic type, cf. Russell [2], Vinter [1], Vinter and Johnson [1].

4. A remark on sensitivity reduction

4.1 Setting of the problem

Let us consider a system whose state equation is again (3.2), (3.3), (3.4) but with a "partly known" operator A . More precisely, let us consider a family $A(\zeta)$ of operators:

$$(4.1) \qquad A(\zeta)\phi = -\Sigma \frac{\partial}{\partial x_i}\left(a_{ij}(x,\zeta)\frac{\partial\phi}{\partial x_j}\right) + \Sigma\, a_j(x,\zeta)\frac{\partial\phi}{\partial x_j} + a_0(x,\zeta)\,\phi$$

where $\zeta \in R$; we suppose that

$$(4.2) \qquad \begin{cases} a_{ij},\ a_j,\ a_0 \in L^\infty(\Omega\times R) \\ \Sigma\, a_{ij}(x,\zeta)\,\eta_i\,\eta_j \geq \alpha\,\Sigma\,\eta_i^2 , \ \alpha>0,\ \forall\zeta \in R . \end{cases}$$

Then <u>for every</u> ζ , the state $y(v,\zeta)$ is the solution of

$$(4.3) \qquad \begin{cases} \frac{\partial y}{\partial t} + A(\zeta)y = f \quad \text{in } Q , \\ \frac{\partial y}{\partial \nu_{A(\zeta)}} = v \quad \text{on } \Sigma , \\ y(0) = y_0 \quad \text{on } \Omega . \end{cases}$$

The <u>cost function</u> is now

$$(4.4) \qquad J(v,\zeta) = \int_\Sigma |y(v,\zeta)-z_d|^2\, d\Sigma + N\int_\Sigma v^2\, d\Sigma .$$

We know that $A(\zeta)$ is "close" to $A(\zeta_0)$, and we would like to obtain an optimal control of "robust" type, i.e. "stable" with respect to changes of $A(\zeta)$ "around" $A(\zeta_0)$. #

A natural idea is to introduce a function $\rho(\zeta)$ such that

$$(4.5) \qquad \begin{cases} \rho \text{ is } \geq 0, \text{ continuous, with compact support around } \zeta_0 , \\ \int_\rho(\zeta)d\zeta = 1 \end{cases}$$

(of course the choice of ρ will depend on the information we have about the system). We now define <u>the cost function</u>

$$(4.6) \qquad J(v) = \int_R \rho(\zeta)d\zeta \int_\Sigma |y(v,\zeta)-z_d|^2\, d\Sigma + N\int_\Sigma v^2\, d\Sigma .$$

The problem we want now to solve is

$$(4.7) \qquad \inf J(v),\ v\in U_{ad}$$

where, as usual, U_{ad} denotes a (non empty) closed convex subset of $L^2(\Sigma)$.

4.2 The optimality system.

Problem (4.7) admits a unique solution u , which is characterized by

$$(4.8)\quad \begin{cases} \int_\rho(\zeta)d\zeta \int_\Sigma (y(u,\zeta)-z_d)\,(y(v,\zeta) - y(u,\zeta))\, d\Sigma + N \int_\Sigma u(v-u)d\Sigma \geq 0 \\ \forall v \in U_{ad},\ u \in U_{ad}\ . \end{cases}$$

Let $p(\zeta)$ be the solution of

$$(4.9)\quad \begin{cases} -\dfrac{\partial p}{\partial t} + A^*(\zeta)\, p = 0 \quad \text{in } Q\ , \\ \dfrac{\partial p}{\partial \nu_{A^*(\zeta)}} = y(\zeta) - z_d \ \text{on } \Sigma,\ (y(\zeta) = y(u,\zeta))\ , \\ p(T) = 0 \quad \text{on } \Omega\ . \end{cases}$$

Then multiplying the first equation (4.9) by $y(v,\zeta) - y(\zeta)$ we obtain

$$0 = - \int_\Sigma (y(\zeta)-z_d)\,(y(v,\zeta)-y(\zeta))\, d\Sigma + \int_\Sigma p(\zeta)\,(v-u)d\Sigma$$

so that (4.8) reduces to

$$(4.10)\quad \int_\Sigma (\int p(\zeta)\, \rho(\zeta)d\zeta + Nu)\,(v-u)\, d\Sigma \geq 0 \ \forall v \in U_{ad},\ u \in U_{ad}\ .$$

Summarizing, <u>the optimality system is given by</u>

$$(4.11)\quad \begin{cases} \dfrac{\partial y}{\partial t} + A(\zeta)y = f,\ -\dfrac{\partial p}{\partial t} + A^*(\zeta)p = 0 \quad \text{in } Q\ , \\ \dfrac{\partial y}{\partial \nu_{A(\zeta)}} = u\ ,\ \dfrac{\partial p}{\partial \nu_{A^*(\zeta)}} = y - z_d \ \text{on } \Sigma\ , \\ y(0) = y_0,\ p(T) = 0 \end{cases}$$

<u>and</u> (4.10).

Remark 4.1

For numerical applications of the preceding remark, as well as for other methods of reduction of sensitivity in the present context, we refer to Abu El Ata [1] and to the bibliography therein.

Remark 4.2

If $\rho(\zeta) \to \delta(\zeta-\zeta_0)$ (= mass + 1 at ζ_0) in the weak star topology of measures (i.e. $\int\rho(\zeta)\,\phi(\zeta)d\zeta \to \phi(\zeta_0)\ \forall\ \phi$ continuous with compact support) and if we denote by u_ρ the solution of (4.7), then one can show that

$$(4.12)\qquad u_\rho \to u \text{ in } L^2(\Sigma) \text{ weakly}$$

where u solves

$$(4.13)\qquad \inf J(v,\zeta_0),\ v\in U_{ad}\ .$$

5. Non well set problems as control problems

5.1 Orientation.

Let us consider the following (non-well-set) problem (this problem arises from a question in medicine, in heart disease; cf. Colli-Franzone, Taccardi and Viganotti [1]): in an open set Ω with boundary $\Gamma_0 \cup \Gamma_1$, a function u satisfies an elliptic equation

$$(5.1)\qquad Ay = 0\ ,$$

and we know that

$$(5.2)\qquad \frac{\partial y}{\partial \nu_A} = 0 \quad \text{on } \Gamma_1$$

and we can measure

$$(5.3)\qquad y = g \quad \text{on } S\ .$$

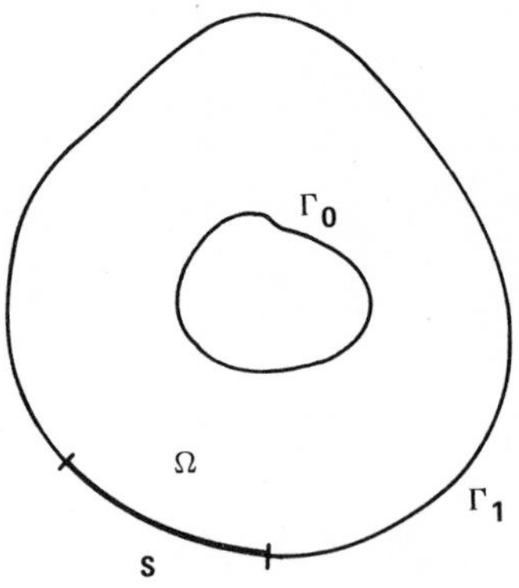

Figure 1

If g is precisely known, this uniquely defines y but, as it is well known, in an unstable manner.

The problem is to find y on Γ_0.

5.2 Formulation as a control problem

Let us define the state $y(v)$ of our system as the solution of

$$(5.4)\qquad \begin{cases} Ay(v) = 0 \text{ in } \Omega , \\ y(v) = v \text{ on } \Gamma_0 , \\ \dfrac{\partial y}{\partial \nu_A}(v) = 0 \text{ on } \Gamma_1 \end{cases}$$

(we assume that the coefficients of A are such that this problem admits a unique solution).

We introduce U_{ad} as the smallest possible closed convex subset of $L^2(\Gamma_0)$ which "contains the information" we may have on the values of y (the "real one") on Γ_0 ; in general it will be of the form

$$(5.5)\qquad \begin{aligned} U_{ad} = \{v|\ v\varepsilon L^2(\Gamma_0),\ m_0(x) \le v(x) \le m_1(x) \text{ on } \Gamma_0 , \\ m_0 \text{ and } m_1 \text{ given in } L^\infty(\Gamma_0)\} . \end{aligned}$$

We introduce the cost function

$$(5.6)\qquad J(v) = \int_S |y(v)-g|^2\, dS$$

and we want to solve the problem

$$(5.7)\qquad \inf J(v),\ v\varepsilon\ U_{ad} .$$

If U_{ad} has been properly chosen (i.e. not too small)

$$(5.8)\qquad \inf J(v) = 0$$

which is attained for $v =$ the value of y on Γ_0.

But, of course, this is again an unstable problem and following the idea of Colli-Franzone, Taccardi and Viganotti, loc. cit. we are now going to regularize the above problem of optimal control.

Remark 5.1

Another approach to the problem stated in 5.1 is given in Lattes-Lions [1] via the Quasi Reversibility method.

5.2 Regularization Method

There are a number of methods available to "stabilize" (5.7). Following Colli-Franzone, et al., we introduce the Sobolev space $H^2(\Gamma_0)$ and the Laplace-Beltrami operator Δ_{Γ_0} on Γ_0. We now consider

$$(5.9) \qquad \tilde{U}_{ad} = \{v \mid v \in H^2(\Gamma_0),\ m_0 \leq v \leq m_1 \text{ on } \Gamma_0\}$$

(we assume that $\tilde{U}_{ad}$ is not empty) and we define, for $\varepsilon > 0$,

$$(5.10) \qquad J_\varepsilon(v) = J(v) + \varepsilon \int_{\Gamma_0} |\Delta_{\Gamma_0} v|^2 \, d\Gamma_0 .$$

The new problem we are considering now is

$$(5.11) \qquad \inf J_\varepsilon(v), \quad v \in U_{ad} .$$

Let u_ε be the unique solution of this problem. The optimality system is given by

$$(5.12) \qquad \begin{cases} Ay = 0,\ A^*p = 0 \text{ in } \Omega , \\ \dfrac{\partial y}{\partial \nu_A} = 0 \text{ on } \Gamma_1 , \\ \dfrac{\partial p}{\partial \nu_{A^*}} = y - g \text{ on } S,\ = 0 \text{ on } \Gamma_1/S , \\ y = u,\ p = 0 \text{ on } \Gamma_0 , \\ \int_{\Gamma_0} \left(- \dfrac{\partial p}{\partial \nu_{A^*}} + \varepsilon \Delta^2_{\Gamma_0} u\right)(v-u)\, d\Gamma_0 \geq 0 \ \forall v \in U_{ad},\ u \in U_{ad} . \end{cases}$$

We refer to the paper of Colli-Franzone et al for the choice of ε and the acutal numerical implementation of this method. For other regularization methods, cf. A. N. Tikhonov [1].

Chapter 2
Duality Methods

1. General considerations.

1.1 Setting of the problem.

Let V and Q be two real Hilbert spaces we consider two functions F and G from V and $Q \to R$ such that

$$(1.1) \quad \begin{cases} F \text{ and } G \text{ are lower semicontinuous and convex on } V \text{ and } \\ Q \text{ respectively, such that} \\ \\ -\infty < F(v) \leq +\infty, \ -\infty < G(q) \leq +\infty \ , \\ F \text{ and } G \text{ are not identically } +\infty \end{cases}$$

(one says that F and G are <u>proper</u> convex functions).

Let be given an operator

$$(1.2) \qquad \Lambda \in L(V;q)$$

and let us consider

$$(1.3) \qquad J(v) = F(v) + G(\Lambda v) \ .$$

We consider the minimization problem

$$(1.4) \qquad \inf J(v), \ v \in V \ .$$

<u>Remark 1.1</u>

Problem (1.4) will be called the <u>primal problem</u> (P). <u>We want to introduce its dual</u> (P*). #

<u>Remark 1.2</u>

By virtue of the fact that F and G can take the value $+\infty$, the formulation (4.3) (4.6) contains - as we will see in the example given in Section 2 - cases <u>with constraints on the control variable</u> v.

<u>Remark 1.3</u>

In the applications what is given is $J(v)$ (and the state equation); one can generally choose F, G, Λ in <u>several different manners</u> - which lead to <u>several different dual problems</u>. #

1.2 A formal computation

We are going to obtain in a formal manner the dual problem (P*). For a justification (under suitable hypotheses on F, G, Λ) cf. T. R. Rockefeller [1], cf. also I. Ekeland and R. Temam [1].

We recall that if Φ is a proper convex function on a Hilbert space H, its conjugate function Φ^* is defined on the dual H^* of H (which can be identified with H or not) as:

$$(1.5)\qquad \begin{cases} \Phi^*(h^*) = \sup\limits_{h} \, [\langle h^*, h\rangle - \Phi(h)] , \\ h^* \in H^*, \ \langle h^*, h\rangle = \text{scalar product in the duality } H^*, H . \end{cases}$$

We introduce

$$(1.6)\qquad \Phi(v,q) = F(v) + G(q)$$

and we remark that

$$(1.7)\qquad \inf_{v} J(v) = \inf_{v}\inf_{q}\sup_{q^*} [\Phi(v,\Lambda v-q) - \langle q^*,q\rangle] ;$$

indeed the sup with respect to q^* is $+\infty$ unless $q = 0$.

If we now commute - in a formal manner - the inf and the sup in (1.7), we obtain

$$(1.8)\qquad \begin{aligned} \inf J(v) &= \sup_{q^*}\inf_{v,q} [\Phi(v, \Lambda v-q] - \langle q^*,q\rangle] \\ &= \sup_{q^*} [-\sup_{v,q} [\langle q^*,q\rangle - \Phi(v,\Lambda v-q]] . \end{aligned}$$

Let us set

$$(1.9)\qquad \Lambda v - q = \zeta .$$

Then

$$\sup_{v,q} [<q^*,q> - \Phi(v,\Lambda v-q)] = \sup [<q^*, \Lambda v-\zeta> - \Phi(v,\zeta)]$$

$$= \sup_{v,\zeta} [<\Lambda^* q^*, v> - F(v) + < -q^*, \zeta> - G(\zeta)] \quad (^1)$$

$$= F^* (\Lambda^* q^*) + G^* (-q^*) ,$$

and we obtain

$$(1.10) \qquad \inf J(v) = \sup_{q^*} [-F^*(\Lambda^* q^*) - G^*(-q^*)] .$$

The problem (P*) is

$$(1.11) \qquad (P^*): \sup_{q^*} [-F^*(\Lambda^* q^*) - G^*(-q^*)] .$$

Remark 1.4

Even if (P) admits a unique solution, say u, (P*) does not necessarily admit a (unique) solution. (cf. Example below in Section 2; other examples are given in Lions [4].)

In case (P*) admits a solution q_0^*, one has

$$(1.12) \qquad F(u) + F^* (\Lambda^* q_0^*) = < u,\Lambda^* q_0^*> ,$$

$$(1.13) \qquad G(\Lambda u) + G^*(-q_0^*) = <\Lambda u, -q_0^* > .$$

2. A problem with constraints on the state.

2.1 Orientation.

We are going to consider a problem whose state equation is a linear parabolic equation and where constraints on the control variable are given through constraints on the state: We show, following an idea of J. Mossino [1], that proper use of duality (as in Section 1) "suppresses" the state constraints at the cost of losing existence of

---

(1) Λ^* is the adjoint of Λ ; $\Lambda^* \in L(Q^*; V^*)$.

a solution of the dual problem. But this procedure gives useful tools for the numerical solution of such problems.

2.2 Setting of the problem.

We consider the state equation (as in Chapter 1, Section 3) given by

(2.1) $$\frac{\partial y}{\partial t} + Ay = 0 \quad \text{in } \Omega \times]0,T[,$$

(2.2) $$\frac{\partial y}{\partial \nu_A} = v \quad \text{on } \Sigma = \Gamma \times]0,T[,$$

(2.3) $$y(0) = 0 ,$$

(We assume the right hand sides of (2.1) (2.3) to be zero, which does not restrict the generality.)

Let $y(v)$ be the solution of (2.1) (2.2) (2.3).

Let y_1 be given in $L^2(\Omega)$. We define

(2.4) $$U_{ad} = \{v \mid v \in L^2(\Sigma),\ y(T;v) = y_1\} ,$$

and we assume that y_1 is such that U_{ad} is not empty. Then U_{ad} is a closed convex subset of $L^2(\Sigma)$.

Let the cost function be given by

(2.5) $$J(v) = \int_\Sigma |y(v)-z_d|^2 \, d\Sigma + N \int_\Sigma v^2 d\Sigma \quad .$$

We consider the problem

(2.6) $$\inf J(v),\ v \in U_{ad} .$$

2.3 Transformation by duality

We define, with the notations of Section 1:

$$V = L^2(\Sigma) ,$$

$$Q = L^2(\Sigma) \times L^2(\Omega) ,$$

$$\Lambda v = \{y(v)|_\Sigma , y(T;v)\} \in Q ,$$

$$F(v) = \frac{N}{2} \int_\Sigma v^2 \, d\Sigma ,$$

$$G(q) = G_1(q_1) + G_2(q_2), \ q = \{q_1, q_2\} , \text{ with}$$

$$G_1(q_1) = \frac{1}{2} \int_\Sigma |q_1 - z_d|^2 \, d\Sigma ,$$

$$G_2(q_2) = \begin{cases} 0 \text{ if } q_2 = y_1 , \\ + \infty \text{ otherwise.} \end{cases}$$

Then (2.6) coincides with (1.4) (with a factor $\frac{1}{2}$). One checks easily that

$$\text{(2.7)} \qquad \begin{cases} F^*(v) = \frac{1}{2N} \int_\Sigma v^2 \, d\Sigma , \quad V^* = V = L^2(\Sigma) , \\ G_1^* (q_1) = \frac{1}{2} \int_\Sigma q_1^2 \, d\Sigma + \int_\Sigma q_1 z_d \, d\Sigma , \ Q^* = Q , \\ G_2^* (q_2) = \int_\Omega q_2 \, y_1 \, dx . \end{cases}$$

For $\{q_1, q_2\} \in Q$, let us define $\Phi(q)$ as the solution of

$$\text{(2.8)} \qquad \begin{cases} - \frac{\partial \Phi}{\partial t} + A^* \Phi = 0 \text{ in } \Omega \times]0,T[, \\ \frac{\partial \Phi}{\partial \nu_{A^*}} = q_1 \text{ on } \Sigma , \\ \Phi(T) = q_2 \text{ on } \Omega . \end{cases}$$

Then one checks that

$$\text{(2.9)} \qquad \Lambda^* q = \Phi(q)|_\Sigma .$$

Indeed, taking the scalar product of the first equation (2.8) with $y(v)$, one obtains

$$0 = \int_{\Omega\times]0,T[} \left(-\frac{\partial\Phi}{\partial t} + A^*\Phi\right) y(v)\, dx\, dt =$$

$$= - \int_\Sigma q_1\, y(v)\, d\Sigma + \int_\Sigma \Phi\, v\, d\Sigma - \int_\Omega \Phi(T)\, y(T,v)\, dx$$

i.e. $\langle q, \Lambda v\rangle = \int_\Sigma \Phi\, v\, d\Sigma$, hence (2.9) follows.

Then according to (1.10)

$$\text{(2.10)} \qquad \inf J(v) = - \inf_q \Big[\frac{1}{2N}\int_\Sigma \Phi(q)^2\, d\Sigma + \frac{1}{2}\int_\Sigma q_1^2\, d\Sigma - \int_\Sigma q_1\, z_d\, d\Sigma - \int_\Omega q_2\, y_1\, dx\Big] ;$$

We see that the dual problem (the $\inf_q$) <u>is a problem without constraints on the "control variable"</u> q ; but it is not coercive in q_2, so that we do not have necessarily existence of a solution of the dual problem; but we have existence of a solution of the regularized dual problem:

$$\text{(2.11)} \qquad \inf_q \Big[\frac{1}{2N}\int_\Sigma \Phi(q)^2\, d\Sigma + \frac{1}{2}\int_\Sigma q_1^2\, d\Sigma + \frac{\varepsilon}{2}\int_\Omega q_2^2\, dx - \int_\Sigma q_1\, z_d\, d\Sigma - \int_\Omega q_2\, y_1\, dx\Big] . \qquad \#$$

<u>Remark 2.1</u>

<u>Optimality system for the regularized dual problem.</u>

Let $q_\varepsilon^0 = \{q_{1\varepsilon}^0, q_{2\varepsilon}^0\}$ be the solution of (2.11). If we set

$$\Phi(q_\varepsilon^0) = \Phi_\varepsilon ,$$

it is characterized by

$$(2.12)\quad \begin{cases} \frac{1}{N} \int_\Sigma \Phi_\varepsilon\,(q)d\Sigma + \int_\Sigma (q^0_{1\varepsilon}-z_d)\ q_1\ d\Sigma \\ \qquad\qquad + \int_\Omega (\varepsilon\ q^0_{2\varepsilon}-y_1)\ q_2\ dx = 0 \\ \forall q\ . \end{cases}$$

We define the "adjoint state" $z\ (=z_\varepsilon)$ by

$$(2.13)\quad \begin{cases} \frac{\partial z}{\partial t} + Az = 0 \quad \text{in} \quad \Omega\times]0,T[\ , \\ \frac{\partial z}{\partial \nu_A} = \frac{1}{N}\Phi_\varepsilon \quad \text{on} \quad \Sigma\ , \\ z(0) = 0\ . \end{cases}$$

Then

$$0 = \int_{\Omega\times]0,T[} \left(\frac{\partial z}{\partial t} + Az\right)\Phi(q)\ dx\ dt = -\frac{1}{N}\int_\Sigma \Phi_\varepsilon\ \Phi(q)\ d\Sigma$$

$$+\int_\Sigma\ z\ q_1\ d\Sigma + \int_\Omega z(T)\ q_2\ dx\ ;$$

therefore (2.12) becomes

$$(2.14)\quad \begin{cases} \int_\Sigma (z+q^0_{1\varepsilon}-z_d)\ q_1\ d\Sigma + \int_\Omega (z(T) + q^0_{2\varepsilon}-y_1)\ q_2\ dx = 0 \\ \forall q\ , \end{cases}$$

and we finally obtain the <u>optimality system</u>

$$(2.15)\quad \begin{cases} -\frac{\partial \Phi_\varepsilon}{\partial t} + A^*\Phi_\varepsilon = 0, \quad \frac{\partial z_\varepsilon}{\partial t} + A\ z_\varepsilon = 0 \quad \text{in} \quad \Omega\times]0,T[\ , \\ \frac{\partial \Phi_\varepsilon}{\partial \nu_{A^*}} = z_d - z_\varepsilon\ , \qquad \frac{\partial z_\varepsilon}{\partial \nu_A} = \frac{1}{N}\Phi_\varepsilon \quad \text{on} \quad \Sigma\ , \\ \Phi_\varepsilon(T) = \frac{1}{\varepsilon}(y_1-z_\varepsilon(T)), \quad z_\varepsilon(0) = 0 \quad \text{on} \quad \Omega\ , \end{cases}$$

with the approximate optimal control u_ε given by

(2.16) $u_\varepsilon + \frac{1}{N}\Phi_\varepsilon$ on Σ . #

2.4 Regularized dual problem and penalized problem

We are going to show, in the setting of the preceding Section, the close connections (actually the identity in this case) which exist between the method of duality and the penalty method.

We consider again the problem (2.5) (2.6) and we define the penalized problem as follows: for $\varepsilon \to 0$, we define

$$J_\varepsilon(v) = J(v) + \frac{1}{\varepsilon}|y(T;v) - y_1|^2 . \tag{2.17}$$

Let u_ε be the unique solution of the penalized problem:

$$J_\varepsilon(u_\varepsilon) = \inf J_\varepsilon(v) , \quad v\varepsilon L^2(\Sigma) . \tag{2.18}$$

Then one shows easily that

$$u_\varepsilon \to u \quad \text{in} \quad L^2(\Sigma) \quad \text{weakly as} \quad \varepsilon \to 0 , \tag{2.19}$$

$$J_\varepsilon(u_\varepsilon) \to J(u\ ) . \tag{2.20}$$

The optimality system is given as follows; we set

$$y_\varepsilon = y(u_\varepsilon)$$

and we define p_ε as the solution of

$$(2.21)\quad \begin{cases} -\dfrac{\partial p_\varepsilon}{\partial t} + A^* p_\varepsilon = 0 \quad \text{in} \quad \Omega\times]0,T[\ , \\ \dfrac{\partial p_\varepsilon}{\partial \nu_{A^*}} = y_\varepsilon - z_d \quad \text{on} \quad \Sigma \ , \\ p_\varepsilon(T) = \dfrac{1}{\varepsilon}\,(y\ (T)-y_1) \quad \text{on} \quad \Omega \ . \end{cases}$$

The optimal control u_ε is characterized by

$$(2.22)\quad \begin{cases} \int_\Sigma (y_\varepsilon - z_d)\ y(v)\ d\Sigma + N \int_\Sigma u_\varepsilon v d\Sigma + \frac{1}{\varepsilon}(y_\varepsilon(T)-y_1,\ y(T;v)) = 0 \\ \forall\ v\in L^2(\Sigma) \ ; \end{cases}$$

using (2.21), (2.22) reduces to $p_\varepsilon + Nu_\varepsilon = 0$ on Σ ; hence the optimality system;

$$(2.23)\quad \begin{cases} \dfrac{\partial y_\varepsilon}{\partial t} + Ay_\varepsilon = 0 \ , \ -\dfrac{\partial p_\varepsilon}{\partial t} + A^* p_\varepsilon = 0 \quad \text{in} \quad \Omega\times]0,T[\ , \\ \dfrac{\partial y_\varepsilon}{\partial \nu_A} + \dfrac{1}{N}\, p_\varepsilon = 0 \ , \ \dfrac{\partial p_\varepsilon}{\partial \nu_{A^*}} = y_\varepsilon - z_d \quad \text{on} \quad \Sigma \ , \\ y_\varepsilon(0) = 0 \ , \ p_\varepsilon(T) = \dfrac{1}{\varepsilon}\,(y_\varepsilon(T)-y_1) \quad \text{on} \quad \Omega \ . \end{cases}$$

This system is actually identical to (2.15) with

$$\Phi_\varepsilon = - p_\varepsilon ,\ z_\varepsilon = Y_\varepsilon .$$

3. Variational principle for the heat equation.

3.1 Direct method

We present here a particular case of a recent result of H. Brézis and I. Ekeland [1].

We want to show that, under suitable hypotheses given below, the solution u of the "heat equation":

$$(3.1)\quad \begin{cases} \frac{\partial u}{\partial t} + Au = f \quad \text{in} \quad \Omega\times]0,T[, \\ u = 0 \quad \text{on} \quad \Sigma , \\ u(x,0) = u_0 \quad \text{in} \quad \Omega \end{cases}$$

realizes the minimum of a quadratic functional.

We assume that

$$(3.2)\qquad A^* = A$$

and that, if $a(\phi,\psi)$ is the linear form associated to A :

$$(3.3)\qquad a(\phi,\phi) \geq \alpha\|\phi\|^2 \ \forall\phi\varepsilon H_0^1(\Omega),\ \alpha > 0 .$$

Remark 3.1

The result below readily extends to higher order elliptic operators A . #

We assume that

$$(3.4)\qquad \begin{array}{l} f\varepsilon L^2(Q) \text{ (actually one could take } f\varepsilon L^2(0,T,H^{-1}(\Omega)) , \\ \text{where } H^{-1}(\Omega) = \text{dual of } H_0^1(\Omega),\ H_0^1(\Omega)\subset L^2(\Omega)\subset H^{-1}(\Omega)) , \\ u_0\varepsilon L^2(\Omega) . \end{array}$$

We define

(3.5) $U = \{\phi \mid \phi \in H^1(\Omega \times]0,T[),\ \phi = 0 \text{ on } \Sigma,\ \phi(x,0) = u_0(x) \text{ on } \Omega\}$.

By virtue of (3.3), A is an <u>isomorphism</u> from $H_0^1(\Omega)$ onto $H^{-1}(\Omega)$, whose inverse is denoted by A^{-1}. We now set

(3.6) $$J(\phi) = \int_0^T [\tfrac{1}{2} a(\phi) + \tfrac{1}{2} a(A^{-1}(f - \tfrac{\partial\phi}{\partial t})) - (f,\phi)]\, dt + \tfrac{1}{2} |\phi(T)|^2$$

where we have used the notation

(3.7) $a(\phi) = a(\phi,\phi)$.

We are going to check that

(3.8) $\inf_{\phi \in U} J(\phi) = J(u)$, u = solution of (3.1) ,

the inf in (3.8) being attained at the unique element u .

<u>Proof</u>: we set

$$\phi = u + \psi ,$$

where ψ spans the set of functions in $H^1(\Omega\times]0,T[)$ such that $\psi = 0$ on Σ , $\psi(0) = 0$.
We have

(3.9) $J(\phi) = J(u) + K(\psi) + X(u,\psi)$,

(3.10) $$K(\psi) = \int_0^T [\tfrac{1}{2} a(\psi) + \tfrac{1}{2} a(A^{-1} \tfrac{\partial\psi}{\partial t})]\, dt + \tfrac{1}{2} |\psi(T)|^2 ,$$

$$X(u,\psi) = \int_0^T [a(u,\psi) - a(A^{-1}(f - \tfrac{\partial u}{\partial t}), A^{-1} \tfrac{\partial\psi}{\partial t}) - (f,\psi)]\, dt + (u(T), \psi(T)) .$$

But from the first equation (3.1) we have

$$A^{-1}\left(f - \frac{\partial u}{\partial t}\right) = u$$

and $a(u, A^{-1}\frac{\partial\psi}{\partial t}) = (u, A\,A^{-1}\frac{\partial\psi}{\partial t}) = (u, \frac{\partial\psi}{\partial t})$, so that

$$X(u,\psi) = \int_0^T [a(u,\psi)-(u, \frac{\partial\psi}{\partial t})-(f,\psi)]dt + (u(T) , \psi(T)) .$$

But taking the scalar product of the first equation (3.1) by ψ gives

$$X(u,\psi) = 0$$

so that

(3.11) $J(\phi) = J(u) + K(\psi)$;

since $K(\psi) \geq 0$ and $K(\psi) = 0$ iff $\psi = 0$, we obtain (3.8) . #

3.2 Use of duality

Let us define

(3.12) $F(\phi) = \frac{1}{2}\,a(\phi)$ on $H_0^1(\Omega)$.

Then the conjugate function F^* of F is given on $H^{-1}(\Omega)$ by

(3.13) $F^*(\phi^*) = \frac{1}{2}\,a(A^{-1}\phi^*)$,

and

(3.14) $$\begin{cases} J(\phi) = \int_0^T [F(\phi)+F^*(f - \frac{\partial\phi}{\partial t}) - \langle\phi, f - \frac{\partial\phi}{\partial t}\rangle]\,dt + \frac{1}{2}|u_0|^2 , \\ \phi\in U . \end{cases}$$

It follows that

$$J(\phi) \geq \frac{1}{2} |u_0|^2$$

and that $J(\phi) = \frac{1}{2} |u_0|^2$ (= J(u)) iff

$$F(\phi) + F^*(f - \frac{\partial\phi}{\partial t}) = \langle \phi, f - \frac{\partial\phi}{\partial t} \rangle$$

i.e.

$$\phi = \text{derivative of } F^* \text{ at } f - \frac{\partial\phi}{\partial t}$$

i.e.

$$\phi = A^{-1} (f - \frac{\partial\phi}{\phi t})$$

i.e.

$$\phi = u .$$

Chapter 3
Asymptotic Methods

1. Orientation

The aim of asymptotic methods in optimal control is to "simplify" the situation by asymptotic expansions of some sort.

This can be achieved by one of the following methods:

(i) simplification of the cost function - this is, for instance the case when the control is "cheap", cf. Section 2;

(ii) simplification of the state equation, by one of the available asymptotic methods:

(j) the most classical one is the use of asymptotic expansions in terms of "small" parameter that may enter the state equation, i.e. the method of perturbations, in particular the method of singular perturbations; we refer for a number of applications in Biochemistry or in Plasma Physics to J. P. Kernevez [1], Brauner and Penel [1], J. Blum [1] and to the bibliography therein; cf. also Lions [7].

(jj) the homogeneization method for operators with highly oscillating coefficients;cf. Section 3;

(jjj) the averaging method of the type of Bogoliubov-Mitropolski [1]; we refer to Bensoussan, Lions and Papanicolaou [1];

(iii) simplification of the "synthesis" operator by the choice of a particular feedback operator (in general on physical grounds); We do not consider this aspect here; we refer to Lions [4], Bermudez [1], Bermudez, Sorine and Yvon [1]; it would be apparently of some interest to consider this question in the framework of perturbation methods.

2. Cheap control. An example.

2.1 Setting of the problem.

With the notations of Chapter 1, Section 3.1, we consider the state equation given by

$$(2.1)\quad \begin{cases} \frac{\partial y}{\partial t} + Ay = f \quad \text{in} \quad Q = \Omega\times]0,T[\ , \\ \frac{\partial y}{\partial \nu_A} = v \quad \text{on} \quad \Sigma \ , \\ y(0) = y_0 \quad \text{on} \quad \Omega \ . \end{cases}$$

We consider the <u>cost function</u>

$$(2.2)\qquad J_\varepsilon(v) = \int_\Sigma |y(v)-z_d|^2 \, d\Sigma + \varepsilon \int_\Sigma v^2 \, d\Sigma$$

where $\varepsilon > 0$ is "small" .

This amounts to considering the control v as "cheap" - a situation which does arise in practical situations, where one often meets the case where acutally $\varepsilon = 0$.

Let U_{ad} be a (non-empty) closed convex subset of $L^2(\Sigma)$, and let u_ε be the solution of

$$(2.3)\quad \begin{cases} J_\varepsilon(u_\varepsilon) = \inf \ J_\varepsilon(v), \ v\varepsilon \ U_{ad} \ , \\ u_\varepsilon \ \varepsilon \ U_{ad} \ . \end{cases}$$

<u>We want to study the behavior</u> of u_ε <u>as</u> $\varepsilon \to 0$.

We shall see that this question is related to problems in <u>singular perturbations</u>.

2.2 <u>A convergence theorem</u>.

Let us set

$$(2.4)\qquad y(u_\varepsilon) = y_\varepsilon \ ;$$

u_ε is characterized by

$$(2.5)\quad \begin{cases} \int_\Sigma (y_\varepsilon - z_d)(y(v)-y_\varepsilon)d\Sigma + \varepsilon \int_\Sigma u_\varepsilon(v-u_\varepsilon)d\Sigma \geq 0, \ \forall v\varepsilon \ U_{ad} \ , \\ u_\varepsilon \ \varepsilon \ U_{ad} \ . \end{cases}$$

We define

$\phi(v) = y(v) - y(0)$ (where here $y(0)$ denotes the solution $y(v)$ of (2.1) for $v = 0$) ; we have

$$(2.6)\quad \begin{cases} \dfrac{\partial\phi}{\partial t}(v) + A\phi(v) = 0, \text{ in } Q , \\ \dfrac{\partial\phi}{\partial\nu_A}(v) = v \text{ on } \Sigma , \\ \phi(v)\big|_{t=0} = 0 \text{ on } \Omega . \end{cases}$$

If we set

$$(2.7)\quad \phi(u_\varepsilon) = \phi_\varepsilon ,$$

(2.5) can be written

$$(2.8)\quad \begin{aligned} &\int_\Sigma \phi_\varepsilon(\phi(v)-\phi_\varepsilon)d\Sigma + \varepsilon \int_\Sigma u_\varepsilon(v-u_\varepsilon)d\Sigma \geq \\ &\geq \int_\Sigma (z_d-y(0))(\phi(v)-\phi_\varepsilon)d\Sigma . \end{aligned}$$

Let us consider the case when

$$(2.9)\quad \Gamma = \partial\Omega \text{ is a } C^\infty \text{ variety ,}$$

and let us write the set Φ of all distributions ϕ in $\Omega\times]0,T[$, which are zero for $t < 0$ and which satisfy

$$(2.10)\quad \frac{\partial\phi}{\partial t} + A\phi = 0 \text{ in } \Omega\times]-\infty ,T[.$$

One can show (cf. Lions-Magenes [1], Vol. 3) that one can define, in a unique manner

$$(2.11)\quad \left\{\phi|_\Sigma , \frac{\partial\phi}{\partial\nu_A}|_\Sigma\right\} \in D'(\Sigma) \times D'(\Sigma) ,$$

$D'(\Sigma)$ = space of distributions on Σ ,

the mapping $\phi \to \left\{\phi|_\Sigma , \frac{\partial\phi}{\partial\nu_A}|_\Sigma\right\}$ being continuous from Φ (provided with the topology of $D'(\Omega\times]-\infty,T[)) \to D'(\Sigma) \times D'(\Sigma)$.

We then define

$$(2.12) \qquad K = \left\{\phi \mid \phi\varepsilon\Phi \ , \ \phi|_\Sigma \ \varepsilon L^2(\Sigma), \ \frac{\partial\phi}{\partial\nu_A}|_\Sigma = M\phi \ \varepsilon L^2(\Sigma)\right\}$$

which is a Hilbert space for the norm

$$(2.13) \qquad (\textstyle\int_\Sigma[\phi^2+(M\phi)^2]d\Sigma)^{1/2} .$$

We define next

$$(2.14) \qquad K_{ad} = \{ \phi \mid \phi\varepsilon K, \ M\phi\varepsilon \ U_{ad} \} ;$$

K_{ad} is a closed convex subset of K .

With these notations, (2.8) is equivalent to

$$(2.15) \qquad \begin{cases} \phi_\varepsilon \ \varepsilon K_{ad} , \\ (\phi_\varepsilon,\phi-\phi_\varepsilon)_{L^2(\Sigma)} + \varepsilon(M\phi_\varepsilon) \ M(\phi-\phi_\varepsilon))_{L^2(\Sigma)} \geq \\ \qquad \geq (z_d-y(0), \ \phi-\phi_\varepsilon)_{L^2(\Sigma)} \quad \forall\phi\varepsilon \ K_{ad} . \end{cases}$$

We can now use general results about singular perturbations in Variational Inequalities; using a result of D. Huet [1], we have:

$$(2.16) \qquad \begin{cases} \phi_\varepsilon \to \phi_0 \ \text{ in } \ L^2(\Sigma) \ \text{ as } \ \varepsilon \to 0 , \\ \text{where } \ \phi_0 \ \text{ is the solution of} \\ (\phi_0,\phi-\phi_0) \geq (z_d-y(0),\phi-\phi_0) \ \forall\phi\varepsilon \ \overline{K_{ad}} , \\ \phi_0\varepsilon\overline{K_{ad}} \end{cases}$$

where

$$(2.17)\qquad \begin{cases} \overline{K_{ad}} = \text{closure of } K_{ad} \text{ in } \hat{K} , \\ \hat{K} = \{\phi \mid \phi\varepsilon\Phi, \phi|_{\Sigma} \ \varepsilon L^2(\Sigma)\} . \end{cases}$$

But if $\text{Proj}_{\overline{K_{ad}}}$ = projection operator in $\hat{K}$ on $\overline{K_{ad}}$, we have

$$(2.18)\qquad \phi_0 = \text{Proj}_{\overline{K_{ad}}} (z_d - y(0))$$

and going back to y_ε one has:

$$(2.19)\qquad y_\varepsilon \to y(0) + \text{Proj}_{\overline{K_{ad}}} (z_d - y(0)) \quad \text{in} \quad L^2(\Sigma) .$$

Remark 2.1

One deduces from (2.19) the convergence of U_ε in a very weak topology.

2.3 Connection with singular perturbations

Consider now the "no-constraints" case. Then (2.18) reduces to

$$(2.20)\qquad \phi_0 = z_d - y(0)$$

so that

$$(2.21)\qquad y_\varepsilon \to z_d \quad \text{in} \quad L^2(\Sigma) .$$

which was easy to obtain directly.

But since in general, considering z_d to be smooth, one does not have $z_d|_{t=0} = y_0|_\Gamma$, the convergence (2.21) cannot be improved (no matter how smooth are the data) in the neighborhood of $t = 0$ on Σ . There is a singular layer around $t = 0$ on Σ .

The computation (in a justified manner) of this type of singular layer is, in general, an open problem.

We refer to Lions [8] for a computation of a surface layer of similar nature, in a simpler situtation, and for other considerations along these lines.

3. Homogeneization

3.1 A model problem

Notation: We consider in R^n functions $y \to a_{ij}(y)$ with the following properties:

$$(3.1) \quad \begin{cases} a_{ij} \in L^\infty(R^n) , \\ a_{ij} \text{ is Y-periodic, i.e. } Y =]0,y_1^0[\times\ldots\times]0,y_n^0[\text{ , and} \\ a_{ij} \text{ is of period } y_k^0 \text{ in the variable } y_k , \\ \Sigma\, a_{ij}(y)\, \zeta_i\zeta_j \geq \alpha\, \Sigma\, \zeta_i^2 \text{ , } \alpha > 0 \text{ , a.e. in } y ; \end{cases}$$

for $\varepsilon>0$, we define the operator A^ε by

$$(3.2) \qquad A^\varepsilon\phi = - \sum_{i,j=1}^{n} \frac{\partial}{\partial x_i} \left(a_{ij}\left(\frac{x}{\varepsilon}\right) \frac{\partial \phi}{\partial x_j}\right) .$$

Remark 3.1

The operator A^ε is a simple case of operators arising in the modelization of composite materials; operators of this type have been the object of study of several recent publications; let us refer to de Giorgi-Spagnolo [1], I. Babuška [1] [2], Bakhbalov [1], Bensoussan-Lions-Papanicolaou [2] and to the bibliography therein.

The state equation

We assume that the state $y_\varepsilon(v)$ is given by

$$(3.3) \qquad \left(\frac{\partial}{\partial t} + A^\varepsilon\right) y_\varepsilon = f \quad \text{in} \quad Q = \Omega\times]0,T[,$$

$$(3.4) \qquad \frac{\partial y}{\partial \nu_{A^\varepsilon}} = v \quad \text{on} \quad \Sigma ,$$

$$(3.5) \qquad y_\varepsilon|_{t=0} = y_0 \quad \text{on} \quad \Omega .$$

The cost function is given by

(3.6) $$J_\varepsilon(v) = \int_\Sigma |y_\varepsilon(v)-z_d|^2\, d\Sigma + N \int_\Sigma v^2 d\Sigma,\ N>0, z_d \in L^2(\Sigma) .$$

Let U_{ad} be a closed convex subset of $L^2(\Sigma)$.

By using Chapter 1, we know that there exists a unique optimal control u_ε , solution of

(3.7) $$J_\varepsilon(u_\varepsilon) = \inf J_\varepsilon(v),\ v \in U_{ad},\ u_\varepsilon \in U_{ad} .$$

The problem we want to study is the behavior of u_ε as $\varepsilon \to 0$.

3.2 The homogeneized operator

Let us consider first the case when v is fixed. One proves then that, when $\varepsilon \to 0$,

(3.8) $$\begin{cases} \frac{\partial y}{\partial t} + Ay = f \text{ in } Q , \\ \frac{\partial y}{\partial \nu_A} = v \text{ on } \Sigma , \\ y|_{t=0} = y_0 \text{ on } \Omega , \end{cases}$$

and where A is given by the following construction.

One defines firstly the operator

(3.9) $$A_1 = -\sum \frac{\partial}{\partial y_i} (a_{ij}(y) \frac{\partial}{\partial y_j}) \text{ on } Y ;$$

for every j one defines X^j as the unique solution, up to an additive constant, of

(3.10) $$\begin{cases} A_1(X^j - y_j) = 0 , \\ X^j \text{ Y-periodic} \end{cases}$$

and one defines next

$$(3.11)\quad \begin{cases} a_{ij} = \frac{1}{|Y|}\, a_1\,(X^j - y_j,\ X^j - y_j), & |Y| = \text{measure of } Y, \\ a_1(\phi,\psi) = \Sigma \int_Y a_{ij}(y)\, \frac{\partial\phi}{\partial y_j}\, \frac{\partial\psi}{\partial y_i}\, dy . \end{cases}$$

Then

$$(3.12)\quad A = -\sum_{i,j=1}^{n} a_{ij}\, \frac{\partial^2}{\partial x_i\, \partial x_j},$$

which defines an elliptic operator with constant coefficients; A is called the homogenized operator associated to A^{ε} .

3.3 A convergence theorem

Let us consider the "homogeneized control problem": let $y(v)$ be defined by (3.8); we define

$$(3.13)\quad J(v) = \int_{\Sigma} |y(v) - z_d|^2\, d\Sigma + N \int_{\Sigma} v^2\, d\Sigma$$

and let u be the unique solution of

$$(3.14)\quad J(u) = \inf\ J(v),\ v\varepsilon U_{ad},\ u\varepsilon\, U_{ad}\ .$$

We have:

$$(3.15)\quad u_{\varepsilon} \to u \quad \text{in} \quad L^2(\Sigma) \quad \text{as} \quad \varepsilon \to 0\ .$$

Proof:

Let us set

$$(3.16)\quad y_{\varepsilon}(u_{\varepsilon}) = y_{\varepsilon},\ y(u) = y\ .$$

Since $J(v) \geq N \int_{\Sigma} v^2 d\Sigma$ we have

$$(3.17)\quad \|u_{\varepsilon}\|_{L^2(\Sigma)} \leq \text{constant}$$

and by virtue of the uniform ellipticity in (3.1), we have

(3.18) $$\|y_\varepsilon\|_{L^2(0,T;H_0^1(\Omega))} \le C ,$$

and also

(3.19) $$\left\|\frac{\partial y_\varepsilon}{\partial t}\right\|_{L^2(0,T;H^{-1}(\Omega))} \le C .$$

It follows from (3.18) (3.19) that

(3.20) $$y_\varepsilon|_\Sigma \quad \varepsilon \text{ compact set of } L^2(\Sigma)$$

and we can extract a subsequence, still denoted by u_ε, y_ε, such that

(3.21) $$u_\varepsilon \to \tilde{u} \text{ in } L^2(\Sigma) \text{ weakly, } \tilde{u} \in U_{ad} ,$$

(3.22) $$\begin{cases} y_\varepsilon \to \tilde{y} \text{ in } L^2(0,T;H_0^1(\Omega)) \text{ weakly,} \\ \dfrac{\partial y_\varepsilon}{\partial t} \to \dfrac{\partial \tilde{y}}{\partial t} \text{ in } L^2(0,T;H^{-1}(\Omega)) \text{ weakly,} \\ y_\varepsilon|_\Sigma \to \tilde{y}|_\Sigma \text{ in } L^2(\Sigma) . \end{cases}$$

Therefore

(3.23) $$\liminf_{\varepsilon \to 0} J_\varepsilon(u_\varepsilon) \ge \int_\Sigma |y-z_d|^2 \, d\Sigma + N \int_\Sigma (\tilde{u})^2 d\Sigma = X .$$

But for every $v \in U_{ad}$, we know that (cf. (3.8)) $y_\varepsilon(v) \to y(v)$ in $L^2(0,T;H_1(\Omega))$ weakly and also that $\frac{\partial y_\varepsilon(v)}{\partial t} \to \frac{\partial}{\partial t} y(v)$ in $L^2(0,T;H^{-1}(\Omega))$ weakly; therefore

(3.24) $$y_\varepsilon(v)|_\Sigma \to y(v)|_\Sigma \text{ in } L^2(\Sigma) \text{ strongly}$$

so that

(3.25) $$J_\varepsilon(v) \to J(v) .$$

Then the inequality $J_\varepsilon(u_\varepsilon) \le J_\varepsilon(v) \forall v \in U_{ad}$ gives

$$X \le J(v), \; v \in U_{ad} \tag{3.26}$$

But one can show that

$$\tilde{y} = y(\tilde{u}) \tag{3.27}$$

so that $X = J(\tilde{u})$, hence (3.26) proves that $\tilde{u} = u$.

Since $\limsup J_\varepsilon(u_\varepsilon) \le J(v) \; \forall v$, we have

$$J_\varepsilon(u_\varepsilon) \to J(u) \; . \tag{3.28}$$

Since $\int_\Sigma |y_\varepsilon - z_d|^2 \, d\Sigma \to \int_\Sigma |y - z_d|^2 \, d\Sigma$ (cf. (3.23)) it follows from (3.28) that

$$N \int_\Sigma u_\varepsilon^2 \, d\Sigma \to N \int_\Sigma u^2 \, d\Sigma \; . \tag{3.29}$$

Since $u_\varepsilon \to u$ in $L^2(\Sigma)$ weakly, it follows from (3.29) that $u_\varepsilon \to u$ in $L^2(\Sigma)$ strongly.

<u>Remark 3.2</u>

Let us consider the optimality system:

$$\begin{cases} \dfrac{\partial y_\varepsilon}{\partial t} + A^\varepsilon y_\varepsilon = f, \; -\dfrac{\partial p_\varepsilon}{\partial t} + (A^\varepsilon)^* p_\varepsilon = 0 \quad \text{in } Q \; , \\[2ex] \dfrac{\partial y_\varepsilon}{\partial \nu_{A^\varepsilon}} = u_\varepsilon, \; \dfrac{\partial p_\varepsilon}{\partial \nu_{(A^\varepsilon)^*}} = y_\varepsilon - z_d \quad \text{on } \Sigma \\[2ex] y_\varepsilon(0) = y_0, \; p_\varepsilon(T) = 0, \text{ on } \Omega \; , \end{cases} \tag{3.30}$$

together with

$$\begin{cases} \int_\Sigma (p_\varepsilon + N u_\varepsilon)(v - u_\varepsilon) \, d\Sigma \ge 0 \; \forall v \in U_{ad} \; , \\[2ex] u_\varepsilon \in U_{ad} \; . \end{cases} \tag{3.31}$$

Then, as $\varepsilon \to 0$,

$$(3.32)\quad \begin{cases} y_\varepsilon \to y \quad \text{in} \quad L^2(0,T;H_0^1(\Omega)) \text{ weakly,} \\ p_\varepsilon \to p \quad \text{in} \quad L^2(0,T;H_0^1(\Omega)) \text{ weakly,} \end{cases}$$

$$(3.33)\quad u_\varepsilon \to u \quad \text{in} \quad L^2(\Sigma) ,$$

where $\{y,p,u\}$ is the solution of the "homogeneized optimality system"

$$(3.34)\quad \begin{cases} \frac{\partial y}{\partial t} + Ay = f, \; -\frac{\partial p}{\partial t} + A^* p = 0 \quad \text{in} \quad Q , \\ \frac{\partial y}{\partial \nu_A} = u, \; \frac{\partial p}{\partial \nu_{A^*}} = y - z_d \quad \text{on} \quad \Sigma , \\ y(0) = y_0, \; p(T) = 0 \quad \text{on} \quad \Omega , \end{cases}$$

with

$$(3.35)\quad \begin{cases} \int_\Sigma (p+Nu)(v-u)\, d\Sigma \geq 0 \;\; \forall v \in U_{ad}, \\ u \in U_{ad} . \end{cases}$$

Remark 3.3

In the "no constraint" case, (3.31) and (3.35) reduce to $p_\varepsilon + Nu_\varepsilon = 0$, $p + Nu = 0$ on Σ .

The optimality system can then be "uncoupled" by the use of a non linear partial differential equation of the Riccati type. The above result leads in this case to an homogeneization result for these non-linear evolution equations.

Chapter 4

Systems Which Are Not of the Linear Quadratic Type

1. State given by eigenvalues or eigenfuncitons.

1.1 Setting of the problem.

Let Ω be a bounded open set in R^n, with a smooth (although this is not indispensable) boundary Γ; Ω is supposed to be connected. Let functions a_{ij} be given in Ω, satisfying

$$\text{(1.1)} \quad \begin{cases} a_{ij} = a_{ji} \in L^\infty(\Omega),\ i.j = 1,\dots,n\ , \\ \Sigma\, a_{ij}(x)\, \zeta_i \zeta_j \geq \alpha\, \Sigma\, \zeta_i^2\ ,\ \alpha > 0\ ,\ \text{a.e. in}\ \Omega\ . \end{cases}$$

Let us consider, as space of controls:

$$\text{(1.2)} \quad U = L^\infty(\Omega)$$

and let us consider U_{ad} such that

$$\text{(1.3)} \quad U_{ad} = \text{bounded closed convex subset of } L^\infty(\Omega)\ .$$

We then consider the eigenvalue problem:

$$\text{(1.4)} \quad \begin{cases} Ay + vy = \lambda y \quad \text{in}\ \Omega\ , \\ y = 0 \quad \text{on}\ \Gamma\ ; \end{cases}$$

it is known (Chicco [1]) that the smallest eigenvalue in (1.4) is simple and that in the corresponding one-dimensional eigen-space there is an eigenfunciton ≥ 0.

We therefore define the state of our system by

$$\text{(1.5)} \quad \{y(v),\ \lambda(v)\}$$

where $\lambda(v)$ = smallest (or first) eigenvalue in (1.4), and

$$\text{(1.6)} \quad \begin{cases} Ay(v) + vy(v) = \lambda(v)\, y(v) \quad \text{in}\ \Omega\ ,\ y(v) = 0 \ \text{on}\ \Gamma \\ y(v) \geq 0 \quad \text{in}\ \Omega\ , \\ |y(v)| = 1 \ (|\cdot| = L^2 \text{norm})\ . \end{cases}$$

The cost function is given by

$$(1.7) \qquad J(v) = \int_\Omega |y(v)-z_d|^2 \, dx ,$$

and the optimization problem we consider consists in finding

$$(1.8) \qquad \inf J(v), \; v \in U_{ad} .$$

1.2 Optimality conditions.

It is a simple matter to see that

$$(1.9) \qquad \begin{cases} v \to \{y(v), \lambda(v)\} & \text{is continuous from } U \text{ weak star} \\ \text{into } H^1(\Omega) \text{ weakly} \times R . \end{cases}$$

Indeed

$$(1.10) \qquad \lambda(v) = \inf_{\phi \in H_0^1(\Omega)} \left[\frac{a(\phi) + \int_\Omega v\phi^2 \, dx}{|\phi|^2} \right]$$

where

$$a(\phi) = a(\phi,\phi)$$

$$a(\phi,\psi) = \Sigma \int_\Omega a_{ij}(x) \frac{\partial\phi}{\partial x_j} \frac{\partial\psi}{\partial x_i} \, dx .$$

Therefore if $v_n \to v$ in $L^\infty(\Omega)$ weak star, it follows from (1.10) that $\lambda(v_n)$ is bounded, hence $y(v_n)$ is bounded in $H_0^1(\Omega)$; we can then extract a subsequence, still denoted by $y(v_n)$, $\lambda(v_n)$ such that $y(v_n) \to y$ in $H_0^1(\Omega)$ weakly and $\lambda(v_n) \to \lambda$. But $y(v_n) \to y$ in $L^2(\Omega)$ strongly, and we have

$$Ay + vy = \lambda y, \; y = 0 \text{ on } \Gamma ,$$

$$y \geq 0, \; |y| = 1$$

so that $y = y(v)$, $\lambda = \lambda(v)$.

It immediately follows from (1.9) that

$$(1.11)\quad \begin{cases} \text{there exists } u \in U_{ad} \ (\underline{\text{not necessarily unique}}) \text{ such that} \\ J(v) = \inf J(v),\ v \in U_{ad}\ . \qquad \# \end{cases}$$

We are now looking for optimality conditions. The main question is of course to study the differentiability of $v \to \{y(v),\ \lambda(v)\}$. Let us make first a formal computation. We set

$$(1.12)\qquad \frac{d}{d\zeta}\, y(v_0 + \zeta v)|_{\zeta=0} = \dot{y},\ \frac{d}{d\zeta}\lambda(v_0 + \zeta v)|_{\zeta=0} = \dot{\lambda}$$

assuming, for the time being, these quantities to exist. Replacing in (1.6) v by $v_0 + \zeta v$ and taking the ζ derivative at the origin, we find

$$A\dot{y} + v_0\, \dot{y} + v\, y(v_0) = \lambda(v_0)\, \dot{y} + \dot{\lambda} y(v_0)$$

i.e.

$$(1.13)\qquad A\dot{y} + v_0\dot{y} - \lambda(v_0)\dot{y} = -vy(v_0) + \dot{\lambda}y(v_0)\ .$$

Of course

$$(1.14)\qquad \dot{y} = 0 \quad \text{on} \quad \Gamma\ .$$

Since $|y(v)| = 1$ we have

$$(1.15)\qquad (\dot{y},\ y(v_0)) = 0\ .$$

Formula (1.10) gives

$$(1.16)\qquad \lambda(v) = a(y(v)) + \int_\Omega v\ y(v)^2\ dx$$

hence

$$(1.17)\qquad \dot{\lambda} = 2a(y(v_0),\ \dot{y}) + 2\int_\Omega v_0\ \dot{y}\ y(v_0)dx + \int_\Omega v\ y(v_0)^2\ dx\ .$$

But from the first equation (1.6) with $v = v_0$ we deduce, by taking the scalar product with $\dot{y}$:

$$a(y(v_0),\dot{y}) + \int_\Omega v_0 y(v_0)\dot{y}\,dx = \lambda(v_0)\int_\Omega y(v_0)\dot{y}\,dx =$$

$$= (\text{by } (1.15)) = 0$$

so that (1.17) gives

(1.18) $$\dot{\lambda} = \int_\Omega v\, y(v_0)^2\,dx .$$

The derivative $\{\dot{y},\dot{\lambda}\}$ is given by (1.13) (1.14) (1.15) (1.18)

Remark 1.1

Since $\lambda(v_0)$ is an eigenvalue of $A + v_0 I$, (1.13) admits a solution iff $(-vy(v_0) + \dot{\lambda}\, y(v_0), y(v_0)) = 0$ which is (1.18).

We can now justify the above calculation:

(1.19) $$\begin{cases} v \to \{y(v), \lambda(v)\} \text{ is Frechet differentiable in } L^\infty(\Omega) \\ \text{with values in } D(A)\times R \end{cases}$$

where

(1.20) $$D(A) = \{\phi \mid \phi\in H_0^1(\Omega),\ A\phi\in L^2(\Omega)\} .$$

This is an application of the implicit function theorem (cf. Mignot [1]); we consider the mapping

(1.21) $$\begin{cases} \phi,\lambda,v \xrightarrow{F} A\phi + v\phi - \lambda\phi \\ D(A)\times R\times U \longrightarrow L^2(\Omega) . \end{cases}$$

This mapping, which is a 2^d degree polynomial, is C^∞ . The partial derivative of F with respect to ϕ,λ at ϕ_0, λ_0, v_0 is given by

(1.22) $$\phi,\lambda \to (A + v_0 - \lambda_0)\phi - \lambda\phi_0 .$$

We consider

S^1 = unit sphere of $L^2(\Omega)$ and we restrict F to $(D(A)\cap S^1)\times R\times U$. If we take in (1.22)

(1.23) $\phi_0 = y(v_0),\ \lambda_0 = \lambda(v_0)$

then (1.22) is an <u>isomorphism</u>; therefore by applying the implicit function theorem, there exists a neighborhood $y\times\Lambda\times U$ of $y(v_0),\lambda(v_0),v_0$ in $(D(A)\cap S^1)\times R\times U$ and there exists a C^∞ function

$$(1.24)\quad \begin{cases} v \longrightarrow \{K_1(v),\ K_2(v)\} \\ U \longrightarrow Y \times \Lambda \end{cases}$$

such that

$$(1.25)\quad \begin{cases} F(K_1(v),\ K_2(v),\ v) = 0,\ v\varepsilon\ U\ , \\ K_1(v_0) = y(v_0),\ K_2(v_0) = \lambda(v_0)\ . \end{cases}$$

We have

$$\frac{\partial F}{\partial\phi}(y(v_0),\lambda(v_0,v_0)\ \dot{y} + \frac{\partial F}{\partial\lambda}(y(v_0),\lambda(v_0),v_0)\dot{\lambda} + \frac{\partial F}{\partial v}(y(v_0),\lambda(v_0),\ v_0) = 0$$

which gives (1.13), hence (1.18) follows and (1.16) (1.15) are immediate. · #

We are now ready to write the <u>optimality conditions</u>: if u is an optimal control then <u>necessarily</u>

(1.26) $(J'(u),\ v-u) \geq 0 \quad \forall v\varepsilon\ U_{ad}$.

We introduce $\dot{y},\ \dot{\lambda}$ with $v-u$ instead of v , and u instead of v_0 , i.e.

$$(1.27)\quad \begin{cases} A\dot{y} + u\dot{y} - \lambda(u)\ \dot{y} = -\ (v-u)\ y(u) + \dot{\lambda}\ y(u)\ , \\ (\dot{y},\ y(u)) = 0\ , \\ \dot{\lambda} = \int_\Omega (v-u)\ y(u)^2\ dx\ , \\ \dot{y} = 0 \ \text{ in } \Gamma\ . \end{cases}$$

Then (1.26) becomes (after dividing by 2), if $y(u) = y$:

(1.28) $\int_\Omega (y-z_d)\, \dot{y}\, dx \geq 0 \quad \forall v \in U_{ad}$.

In order to transform (1.28) we introduce an adjoint state $\{p,\mu\}$ such that

(1.29) $$\begin{cases} Ap + u\, p - \lambda(u)p = y-z_d + \mu y \ , \\ p = 0 \quad \text{on} \quad \Gamma \ ; \end{cases}$$

(1.29) admits a solution iff

(1.30) $(1+\mu)\ |y|^2 = (y,z_d)$.

We uniquely define p by adding the condition

(1.31) $(p,y) = 0$.

Then taking the scalar product of (1.29) with $\dot{y}$, and since $(\dot{y},y) = 0$, we obtain

$$\int_\Omega (y-z_d)\, \dot{y}\, dx = Ap+up-\lambda(u)p,\dot{y}) =$$

$$= (p,A\dot{y}+u\dot{y}-\lambda(u)\dot{y}) = (p,(v-u)y) + \dot{\lambda}(p,y) =$$

$$= -\ (p,(v-u)y)$$

so that we finally obtain <u>the optimality system</u>: <u>in order for</u> u <u>to be an optimal control it is necessary that it satisfies the following system, where</u> $y(u) = y$:

(1.32) $$\begin{cases} Ay + uy = \lambda(u)y, \ y \geq 0, \ |y| = 1 \ , \\ Ap + up - \lambda(u)p = y(y,z_d) - z_d, \ (p,y) = 0, \\ y,\ p = 0 \quad \text{on} \quad \Gamma \end{cases}$$

<u>and</u>

$$(1.33)\qquad \begin{cases} -\int_\Omega py\,(v-u)\,dx \geq 0 & \forall v \varepsilon\, U_{ad}, \\ u \varepsilon\, U_{ad}\ . \end{cases}$$

Let us also remark that the system (1.32) (1.33) admits a solution.

1.3 An example.

The following result is due to Van de Wiele [1]. We consider the case:

$$(1.34)\qquad U_{ad} = \{v \mid k_0 \leq v \leq k_1 \text{ a.e.}\},\ k_i \varepsilon R\ .$$

Then (1.33) is equivaleant to:

$$(1.35)\qquad \begin{cases} py \geq 0 \text{ if } x\, \varepsilon \Omega_1 \Longleftrightarrow u(x) = k_1, \\ py \leq 0 \text{ if } x \varepsilon\, \Omega_0 \Longleftrightarrow u(x) = k_0 \\ py = 0 \text{ in } \Omega \backslash (\Omega_0 \cup \Omega_1) \end{cases}$$

(the sets Ω_0, Ω_1 are defined up to a set of measure 0 ([1])). But it is known that - for a 2^d order elliptic operator - $y(x) > 0$ a.e. so that (1.36) actually reduces to

$$(1.36)\qquad \begin{cases} p \geq 0 \text{ on } \Omega_1, \\ p \leq 0 \text{ on } \Omega_0, \\ p = 0 \text{ on } \Omega\ \ (\Omega_0 \cup \Omega_1)\ . \end{cases}$$

We are going to conclude from this result that

(1.37) if z_d is not an eigenfunction for $A+uI$, and if u is any optimal control, then necessarily ess sup $u = k_1$, ess inf $u = k_0$.

([1]) One can define more precisely these sets up to a set of capacity 0 .

Proof. Suppose on the contrary that, for instance, ess sup $u < k_1$. Then one can find $k > 0$ such that

$$k_0 < u + k < k_1 \tag{1.38}$$

But $y(u+k) = y(u)$, $\lambda(u+k) = \lambda(u)+k$ and $u+k$ is again an optimal control; we have therefore similar conditions to (1.36), but now, by virtue of (1.38), the analogs of Ω_0 and Ω_1 are empty and therefore $p(u+k) = 0$ in Ω , i.e. (cf. (1.32) $y(y,z_d) = z_d$, a case which is excluded. Therefore ess sup $u = k_1$.

2. Another example of a system whose state is given by eigenvalues or eigenfunctions.

2.1 Orientation

We give now another example, arising in the operation of a nuclear reactor. For a more complete study of the example to follow, together with numerical computations, we refer to F. Mignot, C. Saguez and Van de Wiele [1].

2.2 Statement of the problem

The operator A is given as in Section 1. We consider

$$U_{ad} = \{v | \ v \in L^\infty(\Omega),\ 0 < k_0 \le v(x) \le k_1 \text{ a.e. in } \Omega\} . \tag{2.1}$$

The state $\{y(v), \lambda(v)\}$ is defined by

$$\begin{cases} Ay(v) = \lambda(v)\, v\, y(v) & \text{in } \Omega , \\ y(v) = 0 & \text{on } \Gamma . \end{cases} \tag{2.2}$$

$$\begin{cases} \lambda(v) = \text{smallest eigenvalue}, \\ y(v) \ge 0 \quad \text{in } \Omega , \end{cases} \tag{2.3}$$

and $y(v)$ is normalized by

$$(y(v),g) = 1, \ g \text{ given in } L^2(\Omega) . \tag{2.4}$$

We set

$$My(v) = \frac{1}{|\Omega|} \int_\Omega y(v)\, dx \tag{2.5}$$

and we define the cost function by

$$(2.6) \qquad J(v) = \int_\Omega |y(v) - My(v)|^2 \, dx \, .$$

We are looking for

$$(2.7) \qquad \inf J(v), \; v \in U_{ad} \, .$$

Remark 2.1

In (2.4) one can take more generally

$$(2.8) \qquad g \in H^{-1}(\Omega) \, .$$

In particular if the dimension equals 1, we can take

$$(2.9) \qquad g = \Sigma \text{ Dirac measures (cf. Saguez [1])} \, . \qquad \#$$

Remark 2.2

The above problem is a very simplified version of the operation of a nuclear plant where $y(v)$ corresponds to the flux of neutrons and where the goal is to obtain as smooth a flux as possible, which explains why the cost function is given by (2.6). #

2.3 Optimality conditions

As in Section 1 we have existence of an optimal control, say u, in general not unique.

We prove, by a similar argument to the one in Section 1, that $v \to y(v), \lambda(v)$ is Frechet differentiable from $U_{ad} \to D(A) \times R$. If we set

$$(2.10) \qquad y(u) = y, \; \lambda(u) = \lambda \, ,$$

$$(2.11) \qquad \begin{cases} \dot{y} = \dfrac{d}{d\zeta} y(u+\zeta(v-u))|_{\zeta=0} \, , \\ \dot{\lambda} = \dfrac{d}{d\zeta} \lambda(u+\zeta(v-u))|_{\zeta=0} \, , \end{cases}$$

we obtain from (2.2)

$$(2.12)\quad \begin{cases} (A-\lambda u)\ \dot{y} = (\dot{\lambda}u+\lambda(v-u))y, \\ \dot{y} = 0 \quad \text{on} \quad \Gamma , \\ (\dot{y},g) = 0, \\ \dot{\lambda} \int_\Omega uy^2\, dx + \lambda \int_\Omega (v-u)y^2\, dx = 0 . \end{cases}$$

The optimality condition is

$$(2.13)\quad (y-My,\ \dot{y}-M(\dot{y})) \geq 0 \quad \forall v \in U_{ad} .$$

But $\qquad (y-My,\ M\dot{y}) = (M(y-My),\dot{y}) = 0$
so that (2.13) reduces to

$$(2.14)\quad (y-My,\dot{y}) \geq 0 \quad \forall v \in U_{ad} .$$

We define the <u>adjoint state</u> $\{p,\mu\}$, by

$$(2.15)\quad \begin{cases} (A-\lambda u)p = y - My + \mu g, \quad p = 0 \quad \text{on} \quad \Gamma , \\ (p,g) = 0 \end{cases}$$

where μ is such that (2.15) admits a solution, i.e.

$$(y-My,y) + \mu(g,y) = 0 \quad \text{i.e.}$$

$$(2.16)\quad \mu = - (y-My,y) .$$

Taking the scalar product of (2.15) with $\dot{y}$ and using the fact that $(g,\dot{y}) = 0$, we have

$$\begin{aligned} (y-My,\dot{y}) &= ((A-\lambda u)p,\dot{y}) = p,(A-\lambda u)\dot{y}) \\ &= ((\dot{\lambda}u+\lambda(v-u))y,p) \ ; \end{aligned}$$

replacing $\dot{\lambda}$ by its value deduced from the last equation in (2.12), we finally obtain

$$(2.17)\quad \begin{cases} \int_\Omega [p - \frac{(p,uy)}{(y,uy)}\, y]\ y\ (v-u)\ dx \geq 0 \quad \forall v \in U_{ad}, \\ \qquad u \in U_{ad} . \qquad \# \end{cases}$$

Therefore, <u>if</u> u <u>is an optimal control, then one has</u>

$$
(2.18)\quad \begin{cases} (A-\lambda u)y = 0, \\ (A-\lambda u)p = y - My - (y-My,y)g, \\ (g,y) = 1,\ (g,p) = 0, \\ y = p = 0 \quad \text{on} \quad \Gamma . \end{cases}
$$

<u>and</u> (2.17). #

We go one step further, by using the structure of U_{ad} given by (2.1). We introduce, as in Section 1.3,

$$\Omega_i = \{|x| \quad u(x) = k_i\}, \quad i = 0, 1,$$

$$\Omega\setminus(\Omega_0 \cup \Omega_1)$$

and we observe that (2.17) is equivalent to

$$y(p - \frac{(p,uy)}{(y,uy)}\, y) \le 0 \quad \text{on} \quad \Omega_1,$$

$$y(p - \frac{(p,uy)}{(y,uy)}\, y) \ge 0 \quad \text{on} \quad \Omega_0,$$

$$y(p - \frac{(p,uy)}{(y,uy)}\, y) = 0 \quad \text{on} \quad \Omega\setminus(\Omega_0 \cup \Omega_1).$$

But since $y > 0$ a.e. this is equivalent to

$$
(2.19)\quad \begin{cases} p - \dfrac{(p,uy)}{(y,uy)}\, y \le 0 \quad \text{on} \quad \Omega_1, \\ p - \dfrac{(p,uy)}{(y,uy)}\, y \ge 0 \quad \text{on} \quad \Omega_0, \\ p - \dfrac{(p,uy)}{(y,uy)}\, y = 0 \quad \text{on} \quad \Omega\setminus(\Omega_0 \cup \Omega_1) . \end{cases} \qquad \#
$$

We deduce from this remark that

$$(2.20)\quad \begin{cases} \text{if } g \in H^{-1}(\Omega) \text{ and } g \notin L^2(\Omega) \text{ (and even if } g \notin H^1(\Omega), \\ g = \text{constant on } \Gamma) \text{ \underline{then} ess sup } u = k_1, \text{ ess inf } u = k_0 . \end{cases}$$

Proof:

Suppose for instance that ess sup $u < k_1$. Then we can find $\sigma > 1$ such that

$$(2.21)\quad k_0 < \sigma u(x) < k_1 \quad \text{a.e. in } \Omega .$$

But $y(\sigma u) = y(u)$, $\lambda(\sigma u) = \sigma\lambda(u)$ so that σu is again an optimal control and therefore one has the analog of (2.19) but this time with Ω_0 and Ω_1 empty; i.e.

$$(2.22)\quad p - \frac{(p,uy)}{(y,uy)}\, y = 0 \quad \text{a.e. in } \Omega .$$

From the first two equations in (2.18), we deduce from (2.22) that

$$(2.23)\quad y - M(y) = (y-My,y)g \quad \text{a.e. in } \Omega$$

hence the result follows, since (2.23) is impossible under the conditions stated on g in (2.20). #

Remark 2.3

All what has been said in Sections 1 and 2 readily extend to other boundary conditions of the self-adjoint type. #

3. Control in the coefficients

3.1 General remarks

We suppose that the state of the system is given by

$$(3.1)\quad -\Sigma \frac{\partial}{\partial x_i}\left(v(x)\frac{\partial y}{\partial x_j}\right) = f \quad \text{in } \Omega,\ f \in L^2(\Omega),$$

$$(3.2)\quad y = 0 \quad \text{on } \Gamma$$

where $v \in U_{ad}$:

$$(3.3)\quad U_{ad} = \{v \mid v \in L^\infty(\Omega),\ 0 < k_0 \le v(x) \le k_1 \ \text{a.e. in } \Omega\} .$$

Of course (3.1) (3.2) admits a unique solution $y(v) \in H_0^1(\Omega)$.

It is generally not true that $v \to y(v)$ is continuous from $L^\infty(\Omega)$ weak star $\to H_0^1(\Omega)$ weakly and there are indeed counter-examples (cf. Murat [1] [2]) showing that, for cost functions of the type

$$(3.4) \qquad J(v) = |y(v)-z_d|^2$$

there does not exist an optimal control.

Remark 3.1

The control appears in the coefficients of highest order in the operator; when the control appears in lower order terms, the situation is much easier. #

Remark 3.2

Problems of optimal control where the control appears in the highest order derivatives are important in many questions; we refer to the book Lurie [1].

Orientation

In what follows we consider a situation (cf. Cea-Malanowski [1]) when $J(v)$ is of a special form, implying continuity of J for the weak star topology.

3.2 An example

We suppose that the cost function is given by

$$(3.5) \qquad J(v) = (f, y(v)) .$$

Remark 3.3

One can add constraints to (3.3), of the type

$$(3.6) \qquad \int_\Omega v^\alpha(x)dx = \text{given}, \quad \alpha \text{ given} > 0 \text{ or } < 0 \text{ integer} .$$

In case $f = 1$, the problem is to find the composition of materials such that the rigidity of the plate is minimum, or maximum if one looks for the sup of $J(v)$, where v is subject to (3.3), and also possibly to a condition of the type (3.6). #

According to Remark 3.3 it is of interest to consider the two problems, respectively studied by Cea-Malanowski [1] and by Klosowicz-Lurie [1]:

(3.7) $\inf J(v),\ v \in U_{ad}$;

(3.8) $\sup J(v),\ v \in U_{ad}$. #

The main point in solving (3.7) is the following:

(3.9) $\begin{cases} v \to J(v) \text{ is lower semicontinuous from } U_{ad} \text{ provided with} \\ \text{the weak-star topology} \to R \ . \end{cases}$

Proof.

Let $v_n \to v$ in the weak star topology of $L^\infty(\Omega)$. We set

(3.10) $y(v_n) = y_n$.

We have

(3.11) $\|y_n\|_{H_0^1(\Omega)} \leq C$.

Therefore we can extract a subsequence, still denoted by y_n , such that

(3.12) $y_n \to y$ in $H_0^1(\Omega)$ weakly,

but in general $y \neq y(v)$. We observe now (and this is where the very special structure of $J(v)$ comes in) that

$$
(3.13) \quad \begin{aligned} J(v_n) - J(v) &= \int_\Omega v_n\ |\text{grad}(y_n - y(v))|^2\, dx \\ &\quad - \int_\Omega (v_n - v)\ |\text{grad } y(v)|^2\, dx \ . \end{aligned}
$$

Indeed, if we compute the right hand side, we obtain

$$(3.14) \qquad \int_\Omega v_n |\text{grad } y_n|^2 \, dx - 2 \int_\Omega v_n \text{ grad } y_n \text{ grad } y(v) \, dx + \int_\Omega v |\text{grad } y(v)|^2 \, dx \; ;$$

but from (3.1) with $v = v_n$, $y = y_n$, we obtain

$$\int_\Omega v_n \text{ grad } y_n \text{ grad } y(v) \, dx = \int_\Omega f \, y(v) \, dx = \int_\Omega v |\text{grad } y(v)|^2 \, dx = J(v)$$

so that (3.14) actually equals $J(v_n) - J(v)$.

Since $v_n \geq 0$ (since v_n, $v \in U_{ad}$) ; it follows from (3.13) that

$$(3.15) \qquad J(v_n) \geq J(v) - \int_\Omega (v_n - v) \, |\text{grad } y(v)|^2 \, dx$$

and since $v_n - v \to 0$ in $L^\infty(\Omega)$ weak star and since $|\text{grad } y(v)|^2$ is a fixed L^1 function, $\int_\Omega (v_n - v) |\text{grad } y(v)|^2 dx \to 0$ and (3.15) implies

$$(3.16) \qquad \liminf J(v_n) \geq J(v)$$

i.e. (3.9). #

It immediately follows from (3.9) that

(3.17) problem (3.7) admits a solution. #

Remark 3.4

We refer to Cea-Malanowski, loc. cit, for further study of problem (3.7), in particular for numerical algorithms.

Remark 3.5

The existence of an optimal solution in problem (3.8) seems to be open; the proof presented in Klosowitz-Lurie loc. cit. does not seem to be complete, but this paper contains very interesting remarks on the necessary conditions satisfied by an optimal control, assuming it exists. #

Remark 3.6

cf. also Barnes [1] (these Proceedings).

4. A problem where the state is given by an eigenvalue with control in the highest order coefficients.

4.1 Setting of the problem

In $L^\infty(\Omega)$ we consider the open set U defined by

$$U = \{v \mid v \in L^\infty(\Omega),\ v \geq c(v) > 0 \text{ a.e. in } \Omega \text{, where } c(v) \text{ depends on } v\} \quad . \tag{4.1}$$

For every $v \in U$ we define the elliptic operator

$$A_v \phi = - \Sigma \frac{\partial}{\partial x_i} (v(x) \frac{\partial \phi}{\partial x_j}) \ . \tag{4.2}$$

Let k be given > 0. We define the state $y(v)$ as the first eigenfunciton of the problem

$$\begin{cases} A_v y(v) = \lambda(v) (v+k) y(v) \quad \text{in } \Omega , \\ y(v) = 0 \quad \text{on } \Gamma \text{ , (we can normalize } y \text{ by } |y(v)| = 1) , \end{cases} \tag{4.3}$$

where $\lambda(v)$ = smallest eigenvalue, i.e.

$$\phi(v) = \inf_{\phi \in H_0^1(\Omega)} \frac{\int_\Omega v |\text{grad}\phi|^2 \, dx}{\int_\Omega (v+k) \phi^2 \, dx} \ . \tag{4.4}$$

We consider the cost function

$$J(v) = \int_\Omega v \, dx \tag{4.5}$$

and we want to minimize $J(v)$ over the set of v's in U subject to the constraint

$$\lambda(v) = \lambda(1) \ . \tag{4.6}$$

Remark 4.1

This problem has been considered in Armand [1], Jouron [1]. In the application to structural mechanics, n=2 , v corresponds to the width of the structure, and we want to minimize the weight for a first eigenvalue fixed, equal to the eigenvalues of the structure with uniform width equal to 1 . #

Remark 4.2

One will find in Jouron, loc. cit, the study of the analogous problem under the added constraint

$$v(x) \geq c > 0, \text{ c fixed.} \tag{4.7}$$ #

4.2 Optimality conditions

We see, as in Section 1, that in the open set U the functions $v \to \{y(v), \lambda(v)\}$ is Frechet differentiable with values in $H_0^1(\Omega) \times R$.

If we set

$$\begin{cases} \dot{y} = \frac{d}{d\zeta} y(u+\zeta v)|_{\zeta=0} \ , \ \dot{\lambda} = \frac{d}{d\zeta} \lambda(u+\zeta v)|_{\zeta=0} \ , \\ y(v) = y, \ \lambda(u) = \lambda \end{cases} \tag{4.8}$$

(u arbitrarily fixed for the time being), we obtain:

$$A_u \dot{y} + A_v y = \lambda(u+k) \dot{y} + \lambda v y + \dot{\lambda}(u+k) y$$

i.e.

$$\begin{cases} (A_u - \lambda(u+k))\dot{y} = \lambda v y - A_v y + \dot{\lambda}(u+k)y \ , \\ \dot{y} = 0 \text{ on } \Gamma \ . \end{cases} \tag{4.9}$$

This is possible iff the right hand side is orthogonal in $L^2(\Omega)$ to y, hence

$$\dot{\lambda} \int_\Omega (u+k)y^2 \, dx = \int_\Omega v[|\text{grad } y|^2 - \lambda y^2] \, dx \ . \tag{4.10}$$ #

If we assume that there exists $u \in U$ which minimizes (4.5) on the set (4.6), then there exists $\zeta \in R$ (Lagrange multiplier) such that

(4.11) $(J'(u),v) + \zeta \dot{\lambda} = 0 \ \forall v$

i.e., using (4.10):

(4.12) $$1 = \frac{1}{\int_\Omega (u+k)y^2 dx} [|\text{grad } y|^2 - \lambda\, y^2] = 0 \ ;$$

in (4.12) $\lambda=\lambda(u)=\lambda(1)$, so that (4.12) can be written

(4.13) $$|\text{grad } y|^2 - \lambda(1)|y|^2 = \text{constant} = c_1 \ .$$

Since $\lambda(1) = \dfrac{\int_\Omega u|\text{grad } y|^2 \, dx}{\int_\Omega (u+k)\, y^2 \, dx}$, we easily find that

(4.14) $$c_1 = \frac{k\lambda(1)}{\int_\Omega u \, dx} \geq 0 \ . \qquad \#$$

We are going to check that, <u>reciprocally</u>;

(4.15) $$\begin{cases} \text{if } u \in U \text{, with } \lambda(u) = \lambda(1) \text{, is such that} \\ y = y(u) \text{ satisfies} \\ |\text{grad } y|^2 - \lambda(1)|y|^2 = c_1 = \text{positive constant ,} \\ \text{then } u \text{ is an optimal control .} \end{cases}$$

<u>Proof</u>:

Let us multiply the equality in (6.15) by (v-u) and integrate over Ω ; we obtain

$$c_1[J(v)-J(u)] = \int_\Omega v|\text{grad } y|^2 \, dx - \lambda(1) \int_\Omega (v+k)|y|^2 \, dx$$

$$- [\int_\Omega u \ |\text{grad } y|^2 \, dx - \lambda(1)\int_\Omega (u+k) \ |y|^2 \, dx] =$$

$$= \int_\Omega v|\text{grad } y|^2 \, dx - \lambda(1) \int_\Omega (v+k) \ |y|^2 \, dx \geq 0 \text{ (by (4.4))} \ .$$

5. <u>Control of free surfaces</u>.

5.1 <u>Variational Inequalities and free surfaces</u>

Let Ω be a bounded open set in R^n and let a bilinear form $a(\phi,\psi)$ be given on $H_0^1(\Omega)$ (<u>to fix ideas</u>) by

$$(5.1)\qquad \begin{cases} a(\phi,\psi) = \Sigma \int_\Omega a_{ij}(x) \dfrac{\partial\phi}{\partial x_j} \dfrac{\partial\psi}{\partial x_i}\, dx + \int_\Omega a_0 \phi\psi dx + \Sigma \int_\Omega a_j \dfrac{\partial\phi}{\partial x_j}\, \psi\, dx. \\ a_0,\ a_{ij}\ \varepsilon L^\infty(\Omega),\ a_j\ \varepsilon L^\infty(\Omega)\ . \end{cases}$$

We assume that

$$(5.2)\qquad a(\phi,\phi) \geq \alpha\|\phi\|^2,\ \alpha > 0,\quad \phi\varepsilon\ H_0^1(\Omega)\ ,$$

where

$$(5.3)\qquad \|\phi\| = \text{norm of } \phi \text{ in } H_0^1(\Omega)\ .$$

Let K be given such that

(5.4) K is a (non-empty) closed convex subset of $H_0^1(\Omega)$.

Then it is known (cf. Lions-Stampacchia [1]) that if f is given in $H^{-1}(\Omega)$, there exists a unique y such that

$$(5.5)\qquad \begin{cases} y\varepsilon K, \\ a(y,\phi-y) \geq (f,\phi-y) \quad \forall\phi\varepsilon K\ ; \end{cases}$$

(5.5) is what is called a Variational Inequality (V.I.). #

Remark 5.1

If we get $y = y(f)$, we have

$$(5.6)\qquad \|y(f_1)-y(f_2)\| \leq c \quad \|f_1-f_2\|_{H^{-1}(\Omega)}\ . \qquad \#$$

Remark 5.2

In the particular case when a is symmetric:

$$(5.7)\qquad a(\phi,\psi) = a(\psi,\phi) \quad \forall\phi,\psi\varepsilon H_0^1(\Omega)$$

then finding y satisfying (5.5) is equivalent to minimizing

$$(5.8)\qquad \frac{1}{2}\, a(\phi,\phi) - (f,\phi) \quad \text{over } K\ ;$$

then the existence and uniqueness of y in (5.5) is immediate. #

Example 5.1

Let us suppose that

$$(5.9)\qquad \begin{cases} K = \{\phi \mid \phi \geq g \quad \text{a.e. in} \quad \Omega\}\ , \ g \text{ given such that} \\ K \quad \text{is not empty} \ . \end{cases}$$

Then one can, at least formally, interpret (5.5) as follows; if we set in general

$$(5.10)\qquad A\phi = -\sum \frac{\partial}{\partial x_i}\left(a_{ij}\frac{\partial \phi}{\partial x_j}\right) + \sum a_j \frac{\partial \phi}{\partial x_j} + a_0\,\phi$$

then y should satisfy

$$(5.11)\qquad \begin{cases} Ay - f \geq 0\ , \\ y - g \geq 0\ , \\ (Ay - f)\,(y-g) = 0 \quad \text{in} \quad \Omega \end{cases}$$

with

$$(5.12)\qquad y = 0 \quad \text{on} \quad \Gamma\ .$$

We can think of this problem as "a Dirichlet problem with an obstacle", the "obstacle" being represented by g .

The contact region is the set where

$$(5.13)\qquad y(x) - g(x) = 0,\ x \in \Omega\ ;$$

outside the contact region we have the usual equation

$$(5.14)\qquad Ay = f$$

where f represents, for instance, the forces.

The boundary of the contact region is a free surface. Formally one has $y = g$ and $\frac{\partial y}{\partial x_i} = \frac{\partial g}{\partial x_i}$ on this surface.

Remark 5.3

For the study of the regularity of the free surface, we refer to Kinderlehrer [1] and to the bibliography therein.

Remark 5.4

For a systematic approach to the transformation of free boundary problems into V.I. of stationary or of evolution type, we refer to C. Baiocchi [1] and to the bibliography therein. #

Remark 5.5

Actually it has been observed by Baiocchi [2] [3] that one can transform the boundary problems arising in infiltration theory into quasi Variational Inequalities (a notion introduced in Bensoussan-Lions [1] [2] for the solution of impulse control problems). #

There are many interesting papers solving free boundary problems by these techniques; cf. Brezis-Stampacchia [1], Duvaut [1], Friedman [1], Torelli [1], Commincioli [1] and the bibliographies of these works. #

5.2 Optimal control of Variational Inequalities

We define the state $y(v)$ of our system as the solution of the V.I. (with the notions of Section 5.1):

$$(5.15)\qquad \begin{cases} y(v)\in K, \\ a(y(v),\ \phi-y(v)) \geq (f+v,\ \phi-y(v)) \quad \forall\phi\in V \ , \end{cases}$$

where

$$(5.16)\qquad v\in U = L^2(\Omega),\quad v = \text{control function.}$$

The cost function is given by

$$(5.17)\qquad J(v) = |y(v)-z_d|^2 + N|v|^2$$

(where $|\phi|$ = norm of ϕ in $L^2(\Omega)$). #

The optimization problem is then:

$$(5.18)\qquad \inf J(v),\ v\in U_{ad} = \text{closed convex subset of } U\ .$$

It is a simple matter to check that

$$(5.19)\qquad \underline{\text{there exists}}\ u\in U_{ad}\ \underline{\text{such that}}\ J(u) = \inf J(v)\ .$$

Remark 5.6

For cases where we have uniqueness of the solution of problems of this type, cf. Lions [6]. #

Remark 5.7

One can think of problem (5.18) as an optimal control related to the control of free surfaces. In this respect a more realistic problem would be to try to find $v \in U_{ad}$ minimizing the "distance" of the free surface (in case K is given by (5.9)); cf. Example 5.1, Section 5.1) to a given surface. This type of question is still largely open. cf. also Section 6.

We assume from now on that K is given by (5.9). It follows from (5.6) that

$$(5.20) \qquad \|y(v_1) - y(v_2)\| \leq c\ |v_1 - v_2|$$

so that, by a result of N. Aronszajn [1] and F. Mignot [2], the function $v \to y(v)$ is "almost everywhere" differentiable (an extension of a result of Rademacher, established for R^n).

We set formally

$$(5.21) \qquad \dot{y} = \frac{d}{d\zeta} y(u+(v-u))|_{\zeta=0}$$

and we set $y(v) = y$; a necessary condition (but this is formal since we do not know if u is a point where y is differentiable; for precise statements, cf. F. Mignot [2]) of optimality is

$$(5.22) \qquad (y-z_d,\dot{y}) + N(u,v-u) \geq 0 \quad \forall v \in U_{ad} .$$

The main point is now to see what $\dot{y}$ looks like. The optimal state y satisfies:

$$(5.23) \qquad \begin{cases} Ay - (f+u) \geq 0, \\ y - g \geq 0, \\ (Ay - (f+u))\ (y-g) = 0 . \end{cases}$$

Let us introduce

$$(5.24)\qquad \begin{cases} Z = \text{set of } x\text{'s in } \Omega \text{ such that} \\ y(x) - g(x) = 0 \quad (Z \text{ is defined up to a set of measure } 0). \end{cases}$$

Then one can show that, at least "essentially":

$$(5.25)\qquad \begin{cases} \dot{y} = 0 \quad \text{on} \quad Z\,, \\ A\dot{y} = v-u \quad \text{on} \quad \Omega\backslash Z \\ \dot{y} = 0 \quad \text{on} \quad \Gamma\,. \end{cases}$$

This leads to the introduction of the adjoint state by

$$(5.26)\qquad \begin{cases} p = 0 \quad \text{on} \quad Z\,, \\ A^* p = y - z_d \quad \text{On} \quad \Omega\backslash Z \\ p = 0 \quad \text{on} \quad \Gamma\,. \end{cases}$$

Then

$$(y-z_d,\ \dot{y}) = (p,\ v-u)$$

so that (5.22) becomes

$$(5.27)\qquad (p+Nu,\ v-u) \geq 0 \quad \forall v \in U_{ad},\ u \in U_{ad}\,.$$

Conclusion: the optimality system is (formally) given by (5.23) (5.26) (5.27). #

Example 5.2

Let us assume that

$$(5.28)\qquad U_{ad} = U\,.$$

Then (5.27) reduces to

$$(5.29)\qquad p + Nu = 0$$

so that the optimality system becomes:

$$
(5.30)\quad \begin{cases} Ay + \frac{1}{N} p - f \geq 0, \\ y - g \geq 0, \\ (Ay + \frac{1}{N} p - f)(y-g) = 0 \quad \text{in} \quad \Omega , \\ p = 0 \quad \text{on} \quad Z \quad \text{(defined in (5.24))} \\ A^* p = y - z_d \quad \text{on} \quad \Omega\setminus Z, \\ y = p = 0 \quad \text{on} \quad \Gamma . \end{cases}
$$

Let us give another form to (5.30). We introduce a bilinear form on $\phi = H_0^1(\Omega) \times H_0^1(\Omega)$ by

$$
(5.31)\qquad A(y,p;\phi,\psi) = a(y,\phi) + \frac{1}{N} a^*(p,\psi) + \frac{1}{N} (p,\phi) - \frac{1}{N} (y,\psi) ,
$$

where

$$
(5.32)\qquad a^* (\phi,\psi) = a(\psi,\phi) .
$$

We observe that

$$
(5.33)\qquad A(y,p;y,p) = a(y,y) + \frac{1}{N} a^*(p,p) \geq c[\|y\|^2 + \|p\|^2] .
$$

Given ϕ in $H_0^1(\Omega)$ we set

$$
(5.34)\qquad Z(\phi-g) = \text{set of } x\text{'s in } \Omega \text{ such that } \phi(x) - g(x) = 0 .
$$

Then (5.30) can be formulated as:

$$
(5.35)\quad \begin{cases} A(y,p;\phi-y,\psi-p) \geq (f,\phi-y) - (\frac{1}{N} z_d,\psi-p) \\ \forall\phi,\psi \in\phi \quad \text{such that} \quad \phi\geq g,\ \psi=0 \quad \text{on} \quad Z(y-g) , \end{cases}
$$

$$
(5.36)\qquad y,p\in\phi,\ y \geq g,\ p = 0 \quad \text{on} \quad Z(y-g) .
$$

This is a quasi-variational inequality. #

5.3 Open questions

5.3.1 Due to Remark 5.5, it could be of some interest to study the optimal control of systems governed by quasi-variational inequalities.

5.3.2 Even after the interesting results of Mignot [2] for the optimal control of stationary V.I., many questions remain to be solved for the control of V.I. of evolution.

5.3.3 Let us give now an interpretation (cf. Bensoussan-Lions [2], [3]) of $y(v)$ when

$$K = \{\phi \mid \phi \leq 0 \text{ on } \Omega\} \tag{5.37}$$

and, to simplify the exposition,

$$a(\phi,\psi) = \frac{1}{2} \int_\Omega \text{grad } \phi \text{ grad } \psi dx + \Sigma \int_\Omega g_j(x) \frac{\partial \phi}{\partial x_j} \psi \, dx + \int_\Omega \alpha \phi \psi dx \;, \tag{5.38}$$

where the g_j's are , say, in $C^1(\bar{\Omega})$ (in order to avoid here any technical difficulty). Then $y(x;v)$, the solution of the corresponding V.I. (5.15), can be given the following interpretation, as the optimal cost of a stopping time problem.

We define the state of a system, say $z_x(t)$, as the solution of the stochastic differential equation:

$$\begin{cases} dz_x(t) = g(z_x(t))dt + dw(t), \\ z_x(0) = x, \; x \varepsilon \Omega \quad , \end{cases} \tag{5.39}$$

where $g(x) = \{g_j(x)\}$, and where $w(t)$ is a normal Wiener process in R^n . In (5.39) we restrict t to be a.s. $\leq \tau_x$ = exit time of Ω .

Let θ be any stopping time, $\leq \tau_x$.

We define the cost function

$$y_x(\theta) = E \int_0^\theta e^{-\alpha t} \, [f(y_x(t)) + v(y_x(t))]dt \; . \tag{5.40}$$

Then (Bensoussan-Lions, loc. cit.)

$$\inf_{\theta \le \tau_x} y_x(\theta) = y(x;v) \tag{5.41}$$

Question: is it possible to obtain a result of the type (5.25) by using (5.41)? #

6. Geometrical control variables.

6.1 General remarks

Geometrical control variables can appear in several different ways. Among them:

(i) the state can be given by a state equation which contains Dirac masses at points which are somewhat at our disposal;

(ii) the control variable can be the domain itself where we compute the state. #

In the direction (i) we confine ourselves here to refer to Lions [1] [3], Saguez [1] for linear systems cf. also Vallee [1]. Another type of problems containing Dirac masses (all these questions are interesting also for practical applications) is considered in Amouroux [1] and Amouroux and Babary [1].

For non-linear systems, this leads to very interesting questions, also about Partial Differential Equations! Problems of this type were mentioned to the author by Komura [1]. We refer to Bamberger [1], Benilan and H. Brezis [1].

In the direction (ii) the first (and actually the main!) difficulty lies in a convenient definition of the domains. If one parametrizes smoothly the boundaries of the admissible domains then at least as far as existence is concerned there are no great difficulties; cf. Lions [9]. The most general results for the largest possible classes of domains seem to be at present those of D. Chenais [1] [2]. An interesting idea for representing the domain is due to J. Cea [1] [2], with an explicit application in J. P. Zolesio [1].

Assuming this done, or assuming everything is smooth, the next step is to obtain necessary or necessary and sufficient conditions of optimality. (Let us remark that the approach of Cea, loc. cit., simultaneously gives conditions of optimality). A systematic account of this type of problem, with several interesting applications, is given in Pironneau [1]; cf. also Pironneau [2], these proceedings, and the bibliography therein. For extensions of the Hadamard's formula (variation of Green's formula in terms of the variation of the domain), cf. Murat-Simon [1], Dervieux-Palmerio [1].

6.2 Open questions

6.2.1 It would be interesting to extend Hadamard's formula to variational inequalities. A (very partial) attempt toward this goal is made in Lions [10].

6.2.2 The following question (which seems very difficult) is motivated by a paper of Nilson and Tsuei [1] (which presents a much more complicated situation. Let us consider a family of surfaces $\Gamma(v)$

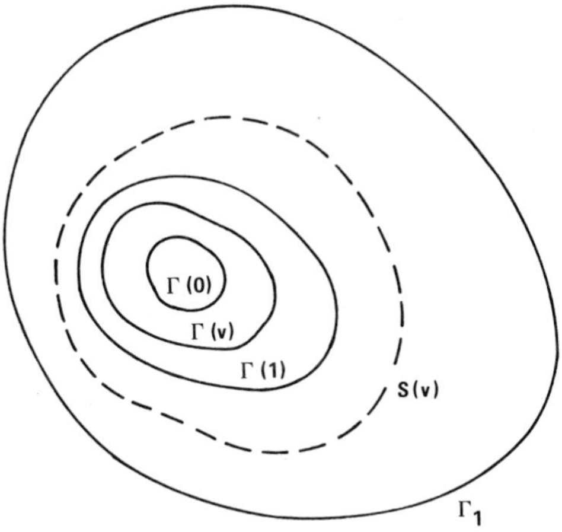

parametrized in some way, where $\Gamma(v)$ lies between $\Gamma(0)$ and $\Gamma(1)$. Let us define $\Omega(v)$ as the open set between $\Gamma(v)$ and a fixed surface Γ_1.

In the domain $\Omega(v)$ we consider the free boundary problems

$$
(6.1) \quad \begin{cases} Ay(v) - f \geq 0, \\ y(v) - g \geq 0, \\ (Ay(v) - f)\,(y(v) - g) = 0 \quad \text{in} \quad \Omega(v) \end{cases}
$$

where f , g are given in $\Omega(0)$ and A is a second order elliptic operator given in $\Omega(0)$; in (6.1) $y(v)$ is subject to some boundary conditions that we do not specify. This V.I. defines a free surface (cf. Section 5.1), denoted by $S(v)$.

The general questions is: what are the surfaces $S(v)$ that one can approximate by allowing $\Gamma(v)$ to be "any" surface between $\Gamma(0)$ and $\Gamma(1)$? (Notice the analogy between this problem and a problem of controllability).

Chapter 5

Remarks on the Numerical Approximation of Problems of Optimal Control

1. General remarks.

Methods for solving numerically problems of optimal control of distributed systems depend on three major possible choices:

(i) choice of the discretization of the state equation (and the adjoint state equation), both in linear and non-linear systems;

(ii) choice of the method to take into account the constraints;

(iii) Choice of the optimization algorithm.

Remark 1.1

If the state is given (as in Chapter 4, Section 1) by the first eigenvalue of course (i) should be replaced by the choice of a method to approximate this first eigenvalue. #

The two main choices for (i) are of course

(i1) finite differences;

(i2) finite elements.

The main trend is now for (i2) and we present below in Section 2 a mixed finite element method which can be used in optimal control.

There are many ways to take into account the constraints, in particular,

(ii1) by duality or Lagrange multipliers;

(ii2) by penalty methods.

Remark 1.2

An interesting method (cf. Glowinski-Marocco [1]) consists in using simultaneously Lagrange multipliers and penalty arguments.

Remark 1.3

One can also consider the state equation, or part of it (such as the boundary conditions) as constraints and use a penalty term for them (cf. Lions [1], Balakrishnan [1], Yvon [3]).

The algorithms used so far for (iii) are:

(iii1) gradient methods in particular in connection with (i1);

(iii2) conjugate gradient methods in particular in connection with (i2);

(iii3) algorithms for finding saddle points such as the Uzawa algorithm.

Remark 1.4

All this is also related to the numerical solution of Variational Inequalities for which we refer to Glowinski, Lions, Trémolières [1].

2. Mixed finite elements and optimal control.

2.1 Mixed variational problems.

We first recall a result of Brezzi [1], which extends a result of Babuška [3]. (cf. also Aziz-Babuška [1].) Let Φ_1, Φ_2 be real Hilbert spaces, provided with the scalar product denoted by $(\, , \,)_i$ (and the corresponding norm being denoted by $\| \, \|_i$, i=1,2) . Let a and b be given bilinear forms:

$$\phi_1,\psi_1 \to a(\phi_1,\psi_1) \text{ is continuous on } \Phi_1 \times \Phi_1 , \tag{2.1}$$

$$\phi_1,\psi_2 \to b(\phi_1,\psi_2) \text{ is continuous on } \Phi_1 \times \Phi_2 . \tag{2.2}$$

We shall assume throughout this section that the following hypothesis hold true: we define

$$\begin{cases} B \in L(\phi_1;\phi'_2) \text{ by} \\ \langle B\phi_1,\psi_2\rangle = b(\phi_1,\psi_2) ; \end{cases} \tag{2.3}$$

we assume

$$a(\phi_1,\phi_1) \geq 0 \quad \forall \phi_1 \in \Phi_1 , \tag{2.4}$$

$$a(\phi_1,\phi_1) \geq \alpha\|\phi_1\|_1^2 , \; \alpha>0, \; \forall\phi_1 \in \text{ Ker } B , \tag{2.5}$$

$$\sup_{\phi_1} \frac{|b(\phi_1,\psi_2)}{\|\phi_1\|_1} \geq c \, \|\psi_2\|_2 , \; c > 0 . \tag{2.6}$$

Remark 2.1

If we introduce $B^* \varepsilon L(\phi_2;\phi_1)$, then (2.6) is equivalent to

$$\text{(2.6)'} \qquad \|B^*\psi_2\|_{\Phi_1'} \geq c\ \|\psi_2\|_2 \forall\psi_2\ \varepsilon\ \Phi_2\ .$$

We now set

$$\text{(2.7)} \qquad \pi(\phi;\psi) = a(\phi_1,\psi_1) + b(\psi_1,\phi_2) - b(\phi_1,\psi_2) \quad \text{on} \quad \Phi \times \Phi$$

where $\Phi = \Phi_1 \times \Phi_2$.

Problem: given a continuous linear form $\psi \to L(\psi)$ on Φ , we look for $\phi\varepsilon\Phi$ such that

$$\text{(2.8)} \qquad \pi(\phi;\psi) = L(\psi)\ \forall\psi\varepsilon\Phi\ .$$

This is what we call a mixed variational problem. For example, we refer to Brezzi, loc cit. and to Bercovier [1]. The result of Brezzi is now:

$$\text{(2.9)} \qquad \begin{cases} \text{under the hypothesis (2.4) (2.5) (2.6) problem (2.8)} \\ \text{admits a unique solution and} \\ \|\phi\|_\Phi \leq c_1\ \|L\|_\Phi\ , \end{cases}$$

The idea of the proof is as follows: we observe first that by virtue of (2.6) B is an isomorphism form $\Phi_1/_{\text{KerB}}$ onto Φ_2' , so that we can find $S\varepsilon L(\Phi'_2;\Phi_1)$ such that $B\cdot S$ = identity.

If we define $A\varepsilon L(\Phi_1;\Phi_1')$ by

$$\text{(2.10)} \qquad <A\phi_1,\psi_1> = a(\phi_1,\psi_1)$$

and if we write

$$\text{(2.11)} \qquad L(\psi) = L_1(\psi_1) + L_2(\psi_2),\ L_i\varepsilon\Phi'_i\ ,$$

then (2.8) is equivalent to

(2.12) $A\phi_1 + B^*\phi_2 = L_1$,

(2.13) $-B\phi_1 = L_2$.

If we set $D_1 = -SL_2$, we have

$$-B(\phi_1 - D_1) = 0 \quad \text{i.e. } z = \phi_1 - D_1 \varepsilon \text{ Ker } B$$

and (2.12) is now equivalent to

(2.14) $Az + B^*\phi_2 = L_1 - AD_1 = g_1$.

But by virtue of (2.5) A is an isomorphism from (Ker B) $\rightarrow$ (Ker B)' ; B is an ismorphism for $\Phi_1/_{\text{Ker B}} \rightarrow \Phi'_2$ so that B^* is an ismorphism from $\Phi_2 \rightarrow (\Phi_1/_{\text{Ker B}})'$. But $\Phi'_1 = (\text{Ker B})' \dotplus (\Phi_1/_{\text{Ker B}})'$; and then $z = A^{-1}h_1$, $\phi_2 = (B^*)^{-1} k_1$.

2.2 Regularization of mixed variational problems

We now follow Bercovier [1].

We define a regularized form $\pi_{\beta,\varepsilon}$ of π by

(2.15)
$$\pi_{\beta,\varepsilon}(\phi;\psi) = \tilde{\pi}(\phi;\psi) = \pi(\phi;\psi) + \beta(\phi_1,\psi_1)_1 + \varepsilon(\phi_2,\psi_2)_2 ,$$
$$\beta,\varepsilon > 0 .$$

We remark that, from (2.4) and (2.7) we have

(2.16) $\pi(\phi;\phi) \geq 0 \ \forall \phi \varepsilon \Phi$

so that

(2.17) $\tilde{\pi}(\phi;\phi) \geq \beta\|\phi_1\|_1^2 + \varepsilon\|\phi_2\|_2^2$.

Therefore there exists a unique element

(2.18) $\phi_{\beta,\varepsilon} = \tilde{\phi} \ \varepsilon \ \Phi$

such that

(2.19) $\tilde{\pi}(\tilde{\phi};\psi) = L(\psi) \ \forall\psi \in \Phi$.

We are now going to prove:

(2.20) $\left\{\begin{array}{l}\text{assuming that one has (2.4) (2.5) (2.6), \underline{and for} } \beta \\ \underline{\text{and}} \ \ \varepsilon/\beta \ \ \underline{\text{small enough, one has}} \\ \|\phi-\tilde{\phi}\|_{\Phi} \leq c(\beta + \frac{\varepsilon}{\beta}) \ \|L\|_{\Phi'} \ .\end{array}\right.$

The proof of (2.20) is in two steps. We introduce

(2.21) $\pi_{\beta}(\phi;\psi) = \hat{\pi}(\phi;\psi) = \pi(\phi;\psi) + \beta(\phi_1,\psi_1)_1$.

We remark that this amounts to replacing $a(\phi_1 \cdot \psi_1)$ by $a(\phi_1,\psi_1) + \beta(\phi_1,\psi_1)_1$ and leaving b invariant. Therefore there exists a unique element $\phi_{\beta} = \hat{\phi}\varepsilon\Phi$ such that

(2.22) $\hat{\pi}(\hat{\phi};\psi) = L(\psi) \ \forall\psi \in \Phi$.

We are going to show

(2.23) $\|\phi-\hat{\phi}\|_{\Phi} \leq C\beta \ \|L\|_{\Phi}$,

(2.24) $\|\hat{\phi}-\tilde{\phi}\|_{\Phi} \leq C \frac{\varepsilon}{\beta} \ \|L\|_{\Phi}$,

from which (2.20) follows.

<u>Proof of</u> (2.23).

We have

(2.25) $\pi(\hat{\phi}-\phi;\psi) + \beta(\hat{\phi}_1,\psi_1)_1 = 0$

so that

$$\hat{\phi}_1-\phi_1 \in \text{Ker } B \ .$$

Therefore if we take $\psi = \hat{\phi} - \phi$ in (2.25) we have (since $\pi(\psi;\psi) \geq \alpha \|\psi_1\|_1^2$ if $\psi_1 \in \operatorname{Ker} B$) :

$$\alpha \|\hat{\phi}_1 - \phi_1\|_1^2 \leq \beta \|\hat{\phi}_1\|_1 \quad \|\hat{\phi}_1 - \phi_1\|_1$$

hence

$$(2.26) \qquad \|\hat{\phi}_1 - \phi_1\|_1 \leq \frac{\beta}{\alpha} \|\hat{\phi}_1\|_1 \ .$$

Therefore

$$\|\hat{\phi}_1\|_1 \leq \|\hat{\phi}_1 - \phi_1\|_1 + \|\phi_1\|_1 \leq \frac{\beta}{\alpha} \|\hat{\phi}_1\|_1 + \|\phi_1\|_1$$

hence, for β small enough,

$\|\hat{\phi}_1\|_1 \leq \frac{1}{1-\beta/\alpha} \|\phi_1\|_1 \leq c \|L\|_{\Phi}$, (we denote by c various constant). Therefore (2.26) implies

$$(2.27) \qquad \|\hat{\phi}_1 - \phi_1\|_1 \leq C\beta \|L\|_{\Phi'} \ .$$

We take now in (2.25) $\psi = \{\psi_1, 0\}$; we notice that

$$(2.28) \qquad \operatorname*{Sup}_{\psi_1} \frac{|\pi(\phi;\psi_1,0)|}{\|\psi_1\|_1} \geq C \|\phi_2\|_2 - C' \|\phi_1\|_1$$

and that (2.25) gives

$$(2.29) \qquad \operatorname{Sup} \frac{|\pi(\hat{\phi}-\phi;\psi_1,0)|}{\|\psi_1\|_1} = \beta \|\hat{\phi}_1\|_1 \ ,$$

so that (2.18) (2.29) give

$$(2.30) \qquad C \|\hat{\phi}_2 - \phi_2\|_2 \leq C' \|\hat{\phi}_1 - \phi_1\|_1 + \beta \|\hat{\phi}_1\|_1 \ ;$$

(2.27) and (2.30), with $\|\hat{\phi}_1\|_1 \leq C\|L\|_{\Phi}$, imply (2.23).

Proof of (2.26)

We have now

$$(2.31) \qquad \pi(\tilde{\phi}-\hat{\phi};\psi) + \beta(\tilde{\phi}_1-\hat{\phi}_1,\psi_1)_1 + \varepsilon(\tilde{\phi}_2,\psi_2)_2 = 0 .$$

Taking $\psi = \tilde{\phi}-\hat{\phi}$ and using (2.16), we obtain

$$(2.32) \qquad \beta\|\tilde{\phi}_1-\hat{\phi}_1\|_1^2 \le \varepsilon\|\hat{\phi}_2\|_2 \, \|\tilde{\phi}_2-\hat{\phi}_2\|_2 .$$

Taking $\psi = \{\psi_1,0\}$ into (2.31) we obtain

$$\text{Sup} \; \frac{|\pi(\tilde{\phi}-\hat{\phi};\psi_1 \;,0)|}{\|\psi_1\|_1} = \beta\|\tilde{\phi}_1 - \phi_1\|_1$$

so that by using (2.28) we obtain

$$(2.33) \qquad \|\tilde{\phi}_2-\hat{\phi}_2\|_2 \le (c'+\beta)\| \tilde{\phi}_1 - \hat{\phi}_1\|_1 .$$

From (2.32) (2.33) we deduce that

$$(2.34) \qquad \|\tilde{\phi}_1-\hat{\phi}_1\|_1 \le C \frac{\varepsilon}{\beta} \|\hat{\phi}_2\|_2 .$$

But with (2.34), (2.33) implies

$$(2.35) \qquad \|\tilde{\phi}_2-\hat{\phi}_2\|_2 \le C \frac{\varepsilon}{\beta} \|\tilde{\phi}_2\|_2$$

and therefore $\|\tilde{\phi}_2\|_2 \le C \frac{\varepsilon}{\beta} \|\tilde{\phi}_2\|_2 + \|\hat{\phi}_2\|_2$ hence it follows that for $\frac{\varepsilon}{\beta}$ small enough

$$\|\tilde{\phi}_2\|_2 \le C \; \|\hat{\phi}_2\|_2$$

and since by (2.23) we have

$$\|\hat{\phi}_2\|_2 \le C \; \|L\|_\Phi \quad ,$$

we have finally

$$(2.36) \qquad \|\tilde{\phi}_2\|_2 \le C \; \|L\|_\Phi \quad .$$

But (2.34) (2.35) (2.36) imply (2.24).

Remark 2.1

If we make a stronger hypothesis than (2.4) (2.5), namely

(2.37) $a\ (\phi_1,\phi_1) \geq \alpha\|\phi_1\|_1^2\ \forall\phi_1\varepsilon\Phi_1$,

then if we introduce

(2.38) $\pi^{\varepsilon}(\phi;\psi) = \pi(\phi;\psi) + \varepsilon(\phi_2,\psi_2)_2$

we have

(2.39) $\pi^{\varepsilon}(\phi;\phi) \geq \alpha\|\phi_1\|_1^2 + \varepsilon\|\phi_2\|_2^2$

and therefore there exists unique $\phi_{\varepsilon}\ \varepsilon\phi$ such that

(2.40) $\pi^{\varepsilon}(\phi_{\varepsilon};\psi) = L(\psi)\ \forall\psi\varepsilon\Phi$.

One has then (cf. Bercovier, loc. cit.)

(2.41) $\|\phi-\phi_{\varepsilon}\|_{\Phi} \leq C\ \varepsilon\ \|L\|_{\Phi}$.

Remark 2.2

Let us define the adjoint form π^* by

(2.42) $\pi^*(\phi;\psi) = \pi(\psi;\phi)$.

If we define a^* by

(2.43) $a^*(\phi_1,\psi_1) = a(\psi_1,\phi_1)$

then

(2.44) $\pi^*(\phi;\psi) = a^*(\phi_1,\psi_1) - b(\psi_1,\phi_2) + b(\phi_1,\psi_2)$.

This amounts to replacing a by a^* and b by $-b$. These changes do not affect (2.4) (2.5) (2.6). We have therefore similar results to the above ones for the adjoint mixed variational problem.

Remark 2.3

Usual variational elliptic problems can be formulated in the preceding setting (cf. Bercovier, loc. cit.); then the approximation results (2.20) or (2.41) lead in a natural way to mixed finite element methods.

2.3 Optimal control of mixed variational systems.

Orientation

We now introduce the standard problems of optimal control for elliptic systems in the mixed variational formulation.

Let U and H be real Hilbert spaces; we are given two operators K and C :

(2.45) $K \in L(U;\Phi')$,

(2.46) $C \in L(\Phi;H)$.

Let π be given by (2.7) and we assume that (2.4) (2.5) (2.6) hold true. Then there exists a unique element $y(v) \in \Phi$ such that

(2.47) $\pi(y(v);\psi) = \langle L+Kv,\psi\rangle \quad \forall\psi\in\Phi$.

This is the state of our system.

The cost function is given by

(2.48) $J(v) = \|Cy(v)-z_d\|_H^2 + N\|v\|_U^2 \ , \ N > 0 \ , \quad z_d \in H$.

Let U_{ad} be a (non-empty) closed convex subset of U .

The optimization problem we want to consider is now

(2.49) $\inf \ J(v), \ v \in U_{ad}$.

Since $v \to y(v)$ is an affine continuous (cf. (2.41)) mapping from $U \to \Phi$, (2.49) admits a unique solution u ; if we set

(2.50) $y(u) = y$,

it is characterized by

$$(2.51)\quad \begin{cases} (Cy-z_d,\ C(y(v)-y))_H + N(u,v-u)_U \geq 0 \ \forall v\varepsilon\ U_{ad}\ , \\ u\varepsilon\ U_{ad}\ . \end{cases}$$

<u>The adjoint state</u>

Using Remark 2.2, one sees that there exists a unique element $p\varepsilon\Phi$ such that

$$(2.52)\quad \pi^*(p;\psi) = (Cy-z_d, C\psi)_H \ \forall\psi\varepsilon\Phi\ ;$$

we call p <u>the adjoint state</u>.

<u>Transformation of</u> (2.51).

By taking $\psi = y(v)-y$ in (2.52) we obtain

$$(2.53)\quad \begin{aligned} (Cy-z_d,\ C(y(v)-y))_H &= \pi^*(p;\ y(v)-y) = \pi(y(v)-y;p) = \\ &= \langle K(v-u), p\rangle\ . \end{aligned}$$

We <u>define</u> K^* by

$$(2.54)\quad (K^*\ p,\ v)_U = \langle Kv,\ p\rangle\ .$$

By virtue of (2.53) (2.54), (2.51) reduces to

$$(2.55)\quad \begin{cases} (K^*p+Nu, v-u)_U \geq 0 \quad \forall v\varepsilon\ U_{ad}\ , \\ u\varepsilon\ U_{ad}\ . \end{cases}$$

<u>The optimality system is finally</u>:

$$(2.56)\quad \begin{cases} \pi(y;\psi) = \langle L + Ku,\ \psi\rangle \ \forall\ \psi\varepsilon\Phi\ , \\ \pi^*(p;\psi) = (Cy-z_d, C\psi)_H \ \forall\ \psi\varepsilon\Phi \end{cases}$$

<u>together with</u> (2.55).

Remark 2.4

Since (2.56) (2.55) is equivalent to the initial problem (2.49) which admits a unique solution, the system (2.56) (2.55) admits a unique solution.

2.4 Approximation of the optimal control of mixed variational systems

We now consider another bilinear form

(2.57) $$\tilde{\pi}(\phi;\psi) = \tilde{a}(\phi_1,\psi_1) + \tilde{b}(\psi_1,\phi_2) - \tilde{b}(\phi_1,\psi_2)$$

with hypothesis on $\tilde{a}$ and $\tilde{b}$ similar to (2.4) (2.5) (2.6). Therefore, with the solutions of Section 2.3, there exists a unique element $\tilde{y}(v)$ such that

(2.58) $$\tilde{\pi}(\tilde{y}(v);\psi) = \langle L + Kv, \psi\rangle \quad \forall\psi\varepsilon\Phi \ .$$

We shall assume that there exists $\rho > 0$ "small" such that

(2.59) $$\|\tilde{y}(v)-y(v)\|_{\Phi} \leq C \ \rho\|v\|_U \ .$$

Remark 2.5

If we take π by (2.15) then, by virtue of (2.20), we have (2.59) with

(2.60) $$\rho = \beta + \frac{\varepsilon}{\beta} \ .$$

Remark 2.6

If we assume (2.37) and (2.6) and if we choose $\tilde{\pi} = \pi^{\varepsilon}$ given by (2.38), then, by virtue of (2.41) we have (2.59) with

(2.61) $$\rho = \varepsilon \ ,$$

(2.62) $$(C\tilde{y}-z_d, C(\tilde{y}(v)-\tilde{y}))_H + N(\tilde{u},v-\tilde{u})_U \geq 0 \quad \forall v\varepsilon \ U_{ad} \ .$$

We have now:

$$(2.63)\quad \begin{cases} \text{if we assume that } \pi \text{ and } \tilde{\pi} \text{ satisfy (2.2) (2.3) (2.4) then} \\ \|u-\tilde{u}\|_U \le C\rho^{1/2} . \end{cases}$$

Remark 2.7

The special hypothesis (2.26) (or an hypothesis on π^*) is needed only for defining the adjoint state; (2.56) is valid without this hypothesis.

Proof of (2.56).

Since $N > 0$, it is enough to consider what happens for a bounded set of v's in U . By virtue of (2.49) we can write

$$(2.64)\quad \tilde{y}(v) = y(v) + r, \ \|r\|_{\Phi} \le C\rho ;$$

therefore

$$\tilde{J}(v) = J(v) + 2(Cy(v)-z_d, Cr)_H + \|Cr\|_H^2$$

so that

$$(2.65)\quad \|\tilde{J}(v) - J(v)\| \le C\rho .$$

We now take $v=\tilde{u}$ in (2.43) and $v=u$ in (2.55). We obtain

$$(Cy-z_d, C(y(\tilde{u}))_H + (C\tilde{y}-z_d, C(\tilde{y}(v)-\tilde{y}))_H - N\|\tilde{u}-u\|_U^2 \ge 0 .$$

We now define

$$(2.66)\quad \tilde{J}(v) = \|C\tilde{y}(v)-z_d\|_H^2 + N\|v\|_U^2$$

and we denote by $\tilde{u}$ the "approximate" optimal control

$$(2.67)\quad \tilde{J}(\tilde{u}) = \inf \tilde{J}(v), \ v\in U_{ad}, \ \tilde{u}\in U_{ad} .$$

It is characterized, if we set $\tilde{y}(\tilde{u}) = \tilde{y}$, by

$$(2.68)\quad \begin{cases} (C\tilde{y}-z_d,\ C(\tilde{y}(v)-\tilde{y}))_H + N(\tilde{u},v-\tilde{u})_H \geq 0 \ \forall v\varepsilon\ U_{ad} , \\ \tilde{u}\varepsilon\ U_{ad} . \end{cases}$$

We now prove the following result:

$$(2.69)\quad \|u-\tilde{u}\|_U \leq C\rho^{1/2} .$$

Proof.

We choose $v = \tilde{u}$ (resp. $v=u$) in (2.51) (resp 2.64)). We obtain

$$(2.70)\quad (C\tilde{y}-z_d,\ C(y(\tilde{u})-y))_H + (C\tilde{y}-z_d,\ C(\tilde{y}(u)-\tilde{y}))_H - N\|\tilde{u}-u\|_U^2 \geq 0 .$$

But (2.66) is equivalent to

$$(2.71)\quad \begin{cases} N\|\tilde{u}-u\|_U^2 + \|C(\tilde{y}-y)\|_H^2 \leq \\ \leq (Cy-z_d,\ \ C(y(\tilde{u})-y)) + (C\tilde{y}-z_d,\ C(\tilde{y}(u)-y)) . \end{cases}$$

Using (2.59) we have

$$\|y(\tilde{u})-\tilde{y}\|_\Phi = \|y(\tilde{u})-\tilde{y}(\tilde{u})\|_\Phi \leq C\rho\|\tilde{u}\|_U,$$

$$\|\tilde{y}(u)-y\|_\Phi = \|\tilde{y}(u)-y(u)\|_\Phi \leq C\rho\|u\|_U$$

so that (2.67) implies

$$(2.72)\quad \|\tilde{u}-u\|_U^2 \leq C\rho(\|u\|_U + \|\tilde{u}\|_U) .$$

But if we choose a fixed $v_0\varepsilon\ U_{ad}$ we have

$$N\|\tilde{u}\|_U^2 \leq \tilde{J}(\tilde{u}) \leq \tilde{J}(v_0) \leq \text{constant}$$

so that (2.68) implies (2.65).

Remark 2.8

We can extend all this theory to the case of evolution equations.

Remark 2.9

For some extensions to non-linear problems, we refer to Bercovier, loc. cit.

Remark 2.10

By using the methods of finite elements for standard elliptic problems (as in Aziz ed: [1], Babuška [1], Brezzi [1], Ciarlet-Raviart [1], Raviart-Thomas [1], Oden [1]) and the above remarks, one obtains in a systematic manner mixed finite element methods for the optimality systems; cf. Bercovier [1].

Remark 2.11

For other approaches, cf. A. Bossavit [1], R. S. Falk [1].

Remark 2.12

We also point out the method of Glowinski-Pironneau [1] who transform non-linear problems in P.D.E. into problems of optimal control, this transformation being very useful from the numerical viewpoint.

Bibliography

1. S. ABU EL ATA [1] Reduction de la sensitivité dans les systèmes Distribués, Thesis, Paris, 1977
2. M. AMOUROUX and J. P. BABARY [1] Optimal pointwise control for a class of distributed parameter systems. IFAC Symp. Boston, August 1975.
3. M. AMOUROUX and J. P. BABARY [1] Determination d'une zone d'action quasi optimale... C.R.A.S. Paris, 1976.
4. J. L. P. ARMAND [1] Application of the theory of optimal control of distributed parameter systems to structural optimization. N.A.S.A. CR 2066, June 1972.
5. N. ARONSZAJN [1] Differentiability of Lipschitzian mappings between Banach spaces. Acta Math., to appear.
6. A. K. Aziz, ed. [1] The Mathematical Foundations of the Finite Element Method. Acad. Press, New York, 1973.
7. I. BABUSKA [1] Reports, University of Maryland, 1976. [2] Homogeneization approach in Engineering. Colloque Versailles 1975. [3] The finite element method with Lagrangian multipliers. Num. Math. 20 (1973), 179-192.

8. C. BAIOCCHI [1] Free boundary problems in the theory of fluid flows through porous media. Proc. Int. C M Vancouver, 1974, Vol. 2, 237-263. [2] Inequations quasi variationnelles dans les problèmes à frontière libre en hydraulique. Colloque IMU-IUTAM. Marseille-September 1975. [3] Studio di un problema quasi variazionale connesso a problemi di frontiera libera. Boll. U.M.I. 1975.
9. N. S. BAKHBALOV [1] Doklady Akad. Nauk. 218 (1974), 1046-1048.
10. A. V. BALAKRISHNAN [1] On a new computing technique in Optimal Control. SIAM J. on Control, (1968), 149-173.
11. A. BAMBERGER [1] To appear.
12. J. S. BARAS and D. G. LAINIOTIS [1] Chandrasekhar algorithms for linear time varying distributed systems. 1976 Conference on Information Sciences and Systems.
13. V. BARBU [1] Constrained control problems with convex costs in Hilbert space. J.M.A.A. 1976.
14. E. R. BARNES [1] These proceedings.
15. D. BEGIS and M. CREPON [1] On the generation of currents by winds: an identification method to determine oceanic parameters. Report Laboria N° 118 - May 1975.
16. Ph. BENILAN and H. BREZIS [1] To appear.
17. A. BENSOUSSAN and J. L. LIONS [1] Notes in the C.R.A.S. Paris on Impulse Control. 276 (1973); 1189-1192; 1333-1338; 278 (1974), 675-579; 747-751. [2] <u>Sur la théorie du Contrôle Optimal</u>. Vol. 1. <u>Temps d'arrêt</u>. Vol. 2 <u>Contrôle impulsionnel</u>. Paris, Hermann, 1977. [3] Problémes de temps d'arrêt optimal et I.V. paraboliques, Applicable Analysis. 3 (1973), 267-295.
18. A. BENSOUSSAN, J. L. LIONS, G. PAPANICOLAOU [1] Book in preparation, North Holland. [2] Notes in the C.R.A.S. Paris, 281 (1975), 89-94; 317-322; 232 (1976), 143-147.
19. M. BERCOVIER [1] Thesis. University of Rouen, 1976.
20. A. BERMUDEZ [1] Contrôle de systèmes distribués, par feedback a priori. Report Laboria, No. 129, June 1975.
21. A. BERMUDEZ, M. SORINE and J. P. YVON [1] To appear.
22. J. BLUM [1] To appear. [2] Identification in Plasma Physics. To appear.
23. N. N. BOGOLIUBOV and Y. A. MITROPOLSKI [1] Asymptotic Methods in the Theory of Nonlinear Oscillation. (Translated from the Russian), Gordon-Breach, 1961.
24. A. BOSSAVIT [1] A linear control problem for a system governed by a partial differential equation. In 9th Int. Conf. on Computing Methods in Optimization Problems - Acad. Press, New York, 1969.
25. J. P. BOUJOT, J. R. MORERA and R. TEMAM [1] An optimal control problem related to the equilibrium of a plasma in a cavity. A. M. and Optimization 2 (1975), 97-129.
26. C. M. BRAUNER [1] Thesis. University of Paris. 1975.
27. C. M. BRAUNER and P. PENEL [1] Un problème de contrôle optimal non linéaire en Biomathématique.Annali Univ. Ferrara, XVII (1973), 1-44. [2] Perturbations singulières...in Lecture Notes in Economics and Math. Systems. Springer 107, 643-668
28. H. BREZIS and I. EKELAND [1]. Un principe variationnnel associé à certaines équations paraboliques. C.R.A.S. 1976.
29. M. BREZIS and G. STAMPACCHIA [1] Annali Scuola Norm. Sup. Pisa, to appear, and C.R.A.S. Paris, 276 (1973), 129-132.

30. F. BREZZI [1] On the existence, uniqueness and approximation of saddle-point problems arising from Lagrangian Multipliers. R.A.I.R.O. (1974), 129-151.
31. A. G. BUTKOVSKY [1] Methods of Control of Distributed Parameter Systems. Moscow, 1975 (in Russian).
32. R. CARROLL [1] Some control problems with differentiably constrained controls. Ric. di Mat. XXIII (1976), 151-157.
33. J. CASTI [1] Matrix Riccati Equations, Dimensionality Reduction. Utilitas Math. 6 (1974), 95-110.
34. J. CASTI and L. LJUNG [1] Some new analytic and computational results for operator Riccati equations. S.I.A.M. J. Control 13 (1975), 817-826.
35. J. CEA [1] Une méthode numérique pour le recherche d'un domaine optimal. Colloquium, Rome, December 1975. [2] Colloquium IRIA, Paris-Versailles, December 1975.
36. J. CEA and K. MALANOWSKI [1] An example of a max-min problem in Partial Differential Equations. SIAM J. Control, 8, (1970), 305-316.
37. G. CHAVENT [1] Identification of distributed parameters. Proc. 3rd IFAC Symp. on Identification, The Hauge, 1973.
38. G. CHAVENT and P. LEMONNIER [1] Estimation des perméabilités relatives... Lecture Notes on Economics and Math. Systems Springer. 107 (1976), p. 440-453.
39. D. CHENAIS [1] On the existence of a solution in a domain identification problem. J.M.A.A. August 1975. [2] To appear.
40. M. CHICCO [1] Some properties of the first eigenvalue and the first eigenfunction of linear second order elliptic partial differential equations in divergence form. Boll. U.M.I. 5 (1972). 245-256.
41. Ph. CIARLET and P. A. RAVIART [1] Mixed finite element methods for the biharmonic equation. In Mathematical Aspects of Finite Elements in P.D.E. Acad. Press, 1974, 125-145.
42. P. COLLI-FRANZONE, B. TACCARDI and C. VIGANOTTI [1] Un metodo per la ricostruzione di potenziali epicardici dai potenziali di superficie. L.A.N. Pavia, 1976.
43. V. COMINCIOLI [1] On some oblique derivative problems...Applied Math and Optimization. Springer, Vol. 1 (1975), 313-336.
44. R. CURTAIN and A. J. PRITCHARD [1] The infinite dimensional Riccati equation for systems defined by evolution operators. Control Theory Centre. Univ. of Warwick. April 1975.
45. M. C. DELFOUR and S. K. MITTER [1] Controllability... of Affine Hereditary Differential Systems. SIAM J. Control (1972), 10, 298-327.
46. A. DERVIEUX, B. PALMERIO [1] Identification de domaines et problèmes de frontières libres. Univ. of Nice, 1974 and C.R.A.S., 1975
47. G. DUVAUT [1] Résolution d'un problème de Stefan. C.R.A.S. Paris, 276 (1973), 1961-1963.
48. I. EKELAND and R. TEMAM [1] Analyse Convexe et Problèmes Variationnels. Paris, Dunod-Gauthier Villars, 1973.

49. R. S. FALK [1] Approximation of a class of optimal control problems with order of convergence estimates. J.M.A.A. 44, (1973), 28-47.
50. H. FATTORINI [1] These proceedings.
51. A. FRIEDMAN [1]
52. E. de GIORGI and S. SPAGNOLO [1] Sulla convergenza degli integrali dell' energia per operatori ellittici del 2° ordine. Boll. U.M.I. 8 (1973), 391-411.
53. R. GLOWINSKI [1] Lagrange and penalty.
54. R. GLOWINSKI, J. L. LIONS and R. TREMOLIERES [1] Analyse Numerique des Inequations Variationelles. Paris, Dunod, 1976.
55. R. GLOWINSKI and I. MAROCCO [1] Sur l' approximation... R.A.I.R.O. (1975), 41-76.
56. R. GLOWINSKI and O. PIRONNEAU [1] Calcul d'ecoulements transoniques. Colloque IRIA-Laboria, Versailles, December 1975.
57. D. HUET [1] Perturbations singulières d'Inegalités Variationnelles. C.R.A.S. 267 (1968), 932-946.
58. C. JOURON [1] Etude des conditions nécessaires k'optimalité pour un problème d'optimisation non convexe. C.R.A.S. Paris 281 (1975). 1031-1034.
59. J. P. KERNEVEZ [1] Control of the flux of substrate entering an enzymatic membrane by an inhibitor at the boundary. J. Optimization Theory and Appl. 1973. [2] Book to appear.
60. D. KINDERLEHRER [1] Lecture at the I.C.M. Vancouver, 1974.
61. B. KLOSOWICZ and K. A. LURIE [1] On the optimal nonhomogeneity of a torsional elastic bar. Archives of Mechanics 24 (1971), 239-249.
62. KOMURA [1] Personal Communication, Tokyo, 1975.
63. C. C. KWAN and K. N.WANG [1] Sur la stabilisation de la vibration elastique. Scientia Sinica, VXII (1974), 446-467.
64. R. LATTES and J. L. LIONS [1] La Methode de Quasi Reversibilité et Applications. Paris, Dunod, 1967. (Elsevier, English translation, by R. Bellman, 1970).
65. J. L. Lions [1] Sur le contrôle optimal des systèmes gouvernés par des equations aux derivées partieles. Paris, Dunod-Gauthier Villars, 1968. (English translation by S. K. Mitter, Springer, 1971.) [2] Equations différentielles opérationnelles et problèmes aux limites. Springer, 1961. [3] Some aspects of the optimal control of distributed parameter systems. Reg. Conf. S. in Appl. Math., SIAM, G, 1972. [4] Various topics in the theory of optimal control of distributed systems, in Lecture Notes in Economics and Math. Systems, Springer, Vol. 105, 1976 (B. J. Kirby, ed.), 166-303. [5] Sur le contrôle optimal de systèmes distribués. Enseignement Mathematique, XIX (1973), 125-166. [6] On variational inequalities (in Russian), Uspekhi Mat. Nauk, XXVI (158), (1971), 206-261. [7] Perturbations singulières dans les problèmes aux limites et en contrôle optimal. Lecture Notes in Math., Springer, 323, 1973. [8] Contrôle optimal de systèmes distribués: propriétés de comparaison et perturbations singulières. Lectures at the Congress :Metodi Valutativi nella Fisica - Mathematica:, Rome, December 1972. Accad. Naz. Lincei, 1975, 17-32. [9] On the optimal control of distributed parameter systems, in Techniques of optimization, ed. by A. V. Balakrishnan, Acad. Press, 1972. [10] Lecture

in Holland.
66. J. L. LIONS and E. MAGENES [1] Problèmes aux limites non homogènes et applications. Paris, Dunod, Vol. 1, 2, 1968; Vol. 3, 1970. English translation by P. Kenneth, Springer, 1972, 1973.
67. J. L. LIONS and G. STAMPACCHIA [1] Variational Inequalities. C.P.A.M. XX (1967), 493-519.
68. K. A. LURIE [1] Optimal control in problems of Mathematical Physics. Moscow, 1975.
69. G. I. MARCHUK [1] Conference IFIP Symp. Optimization, Nice, September 1975.
70. F. MIGNOT [1] Contrôle de fonction propre. C.R.A.S. Paris, 280 (1975), 333-335. [2] Contrôle dans les Inequations Elliptiques. J. Functional Analysis. 1976.
71. F. MIGNOT, C. SAGUEZ and J. P. VAN DE WIELE [1] Contrôle Optimal de systèmes gouvernés par des problèmes aux valeurs propres. Report Laboria, 1976.
72. J. MOSSINO [1] An application of duality to distributed optimal control problems...J.M.A.A. (1975), 50, p. 223-242. [2] A numerical approach for optimal control problems...Calcolo (1976).
73. F. MURAT [1] Un contre exemple pour le problème du contrôle dans les coefficients. C.R.A.S. 273 (1971), 708-711. [2] Contre exemples pour divers problèmes où le contrôle intervient dans les coefficients. Annali M. P. ed. Appl. 1976.
74. F. MURAT and J. SIMON [1] To appear.
75. R. H. NILSON and Y. G. TSUEI [1] Free boundary problem of ECM by alternating field technique on inverted plane. Computer Methods in Applied Mech. and Eng. 6 (1975), 265-282.
76. J. T. ODEN [1] Generalized conjugate functions for mixed finite element approximations..., in The Mathematical Foundations of the Finite Element Method, A. K. Aziz, ed., 629-670, Acad. Press, New York, 1973.
77. O. PIRONNEAU [1] Sur les problèmes d'optimisation de structure en Mécanique des fluides. Thesis, Paris, 1976. [2] These proceedings.
78. M. P. POLIS and R. E. GOODSON [1] Proc. I.E.E.E., 64(1976), 45-61.
79. P. A. RAVIART and J. M. THOMAS [1] Mixed finite elements for 2nd order elliptic problems. Conf. Rome, 1975.
80. W. H. RAY and D. G. LAINIOTIS, ed. [1] Identification, Estimation and Control of Distributed Parameter Systems.
81. R. T. ROCKAFELLAR [1] Conjugate duality and optimization. Reg. Conf. Series in Applied Math. SIAM. 16, 1974.
82. D. L. RUSSELL [1] These proceedings. [2] Control theory of hyperbolic equations related to certain questions in harmonic analysis and spectral theory. J.M.A.A. 40 (1972), 336-368.
83. C. SAGUEZ [1] Integer programming applied to optimal control. Int. Conf. Op. Research, Eger. Hungary, August 1974.
84. J. SAINT JEAN PAULIN [1] Contrôle en cascade dans un problème de transmission. To appear.
85. Y. SAKAWA and T. MATSUSHITA [1] Feedback stabilization of a class of distributed systems and construction of a state estimator. IEEE Transactions on Automatic Control, AC-20, 1975, 748-753.
86. J. SUNG and C. Y. YÜ [1] On the theory of distributed parameter systems with ordinary feedback control. Scientia Sinica, SVIII, (1975), 281-310.

87. L. TARTAR [1] Sur l'étude directe d'equations non lineaires intervenant en théorie du contrôle optimal. J. Funct. Analysis 17 (1974),1-47. [2] To appear.
88. A. N. TIKHONOV [1] The regularization of incorrectly posed problems. Doklady Akad. Nauk SSSR,153 (1963), 51-52, (Soviet Math. 4, 1963, 1624-1625).
89. G. TORELLI [1] On a free boundary value problem connected with a nonsteady filtration phenomenon. To appear.
90. A. VALLEE [1] Un problème de contrôle optimum dans certains problèmes d'evolution. Ann. Sc. Norm Sup. Pisa, 20 (1966), 25-30.
91. J. P. VAN DE WIELE [1] Résolution numerique d'un problème de contrôle optimal de valeurs propres et vecteurs propres. Thesis 3rd Cycle. Paris 1976.
92. R. B. VINTER [1] Optimal control of non-symmetric hyperbolic systems in n-variables on the half space. Imperial College Rep. 1974.
93. R. B. VINTER and T. L. JOHNSON [1] Optimal control of non-symmetric hyperbolic systems in n variables on the half-space. To appear.
94. P. K. C. WANG [1].
95. J. L. A. YEBRA [1]. To appear.
96. J. P. YVON [1] Some optimal control problems for distributed systems and their numerical solutions. [2] Contrôle optimal d'un problème de fusion. Calcolo. [3] Etude de la methode de boucle ouverte adaptee pour le contrôle de systèmes distribués. Lecture Notes in Economics and Math. Systems, 107, (1974), 427-439. [4] Optimal control of systems governed by V.I. Lecture Notes in Computer Sciences, Springer, 3 (1973), 265-275.
97. J. P. ZOLESIO [1] Univ. of Nice Report, 1976.

We also refer to:
Report of the Laboratoire d'Automatique, E.N.S. Mecanique, Nantes: Calculateur hybride et Systèmes a paramètres répartis, 1975

"STOCHASTIC FILTERING AND CONTROL OF LINEAR SYSTEMS: A GENERAL THEORY"

A. V. Balakrishnan*

A large class of filtering and control problems for linear systems can be described as follows. We have an observed (stochastic) process $y(t)$ (say, an $m \times 1$ vector), t representing continuous time, $0 < t < T < \infty$. This process has the structure:

$$y(t) = v(t) + n_0(t)$$

where $n_0(t)$ is the unavoidable measurement error modelled as a white Gaussian noise process of known spectral density matrix, taken as the Identity matrix for simplicity of notation. The output $v(t)$ is composed of two parts: the response to the control input $u(t)$ and a random 'disturbance' $n_L(t)$ (sometimes referred to as 'load disturbance' or 'stale noise') also modelled as stationary Gaussian; we also assume the system responding to the control is linear and time-invariant so that we have:

$$v(t) = \int_0^t B(t-s)\, u(s)ds + n_L(t)$$

where $u(\cdot)$ is always assumed to be locally square integrable, and where $B(\cdot)$ is a 'rectangular' matrix function and

$$\int_0^\infty ||B(t)||^2 dt < \infty \ .$$

* Research supported in part under grant no. 73-2492, Applied Mathematics Division, AFOSR, USAF

We assume further more that the random disturbance is 'physically realizable' so that we can exploit the representation:

$$n_L(t) = \int_0^t F(t-\rho)\ N(\rho)\ d\rho$$

where $F(\rho)$ is a rectangular matrix such that

$$\int_0^\infty ||F(s)||^2 ds < \infty$$

where, in the usual notation,

$$||A||^2 = \text{Tr.}\ \ AA^* .$$

We assume that the process $n_L(t)$ is independent of the observation noise process $n_0(t)$.

It is more convenient now to rewrite the total representation as:

$$\left.\begin{aligned} y(t,\omega) &= v(t,\omega) + G\omega(t) \\ v(t,\omega) &= \int_0^t B(t-s)\ u(s)ds + \int_0^t F(t-s)\ \omega(s)ds \end{aligned}\right\} \tag{1.1}$$

where

$$GG^* = I$$

$$F(t)G^* = 0$$

$\omega(\cdot)$ is white noise process in the appropriate product Euclidean space, and

$$\int_0^\infty ||F(t)||^2 dt < \infty .$$

We hasten to point out that we may replace the white noise formalism by a 'Wiener process' formalism for the above as:

$$Y(t,\omega) = \int_0^t v(s,\omega)ds + G\ W(t,\omega)$$

$$v(t,\omega) = \int_0^t B(t-s)u(s)ds + \int_0^t F(t-s)dW(s,\omega)$$

It makes no difference to the theory that follows as to which formalism is used. The optimization problem we shall consider is a stochastic control ("regulator") problem in which the filtering problem is implicit: to minimize the effect of the disturbance on the output (or some components of it). More specifically, we wish to minimize:

$$E \int_0^t [Qv(t,\omega), Qv(t,\omega)]dt + E \int_0^t [u(t,\omega), u(t,\omega)]dt , \tag{1.2}$$

E denoting expectation, where for each t, $u(t,\omega)$ must 'depend' only upon the available observation up to time t. We can show [1] that under the representation (1.1), (1.2), the optimal control may be sought in the class of 'linear' controls of the form:

$$u(t,\omega) = \int_0^t K(t,s)dY(s,\omega)$$

in the Wiener process formalism, or

$$\int_0^t K(t,s)\, y(s,\omega)ds \tag{1.3}$$

in the white noise formalism.

This problem embraces already all the stochastic control problems for systems governed by ordinary differential equations by taking the special case where the Laplace transforms of $B(\cdot)$ and $F(\cdot)$ are rational. But it also includes a wide variety of problems involving partial differential equations where the observation process $Y(t)$ for each t has its range in a finite dimensional Euclidean space (measurements at a finite number of points in the domain or on the boundary for example). One may argue that any physical measurement must be finite dimensional; in any case, the extension to the infinite dimensional case brings little that is new, and we shall not go into it here.

As a simple example of a non-rational case we may mention:

$$F(t) = t^{-3/2}\, e^{-1/t} \tag{1.4}$$

arising from boundary input in a half-infinite rod [5]. Note that the associated process $n_L(t)$ is not 'Markovian' even in the extended sense [2].

To solve our problem, our basic technique is to create an 'artificial' state space representation for (1.1). It is artificial in the sense that it has nothing to do with the actual state space that originates with the problems. We shall illustrate this with a specific example below. Without going into the system theoretic aspects involved, let us simply note that the controllable part of the original state space can be put in one-to-one correspondence with the controllable part of the artificial state space.

Let H denote $L_2[0,\infty;R_m]$ where m is the dimension of the observation process. Let A denote the operator with domain in H:

$\mathcal{D}(A) = [f \varepsilon H \mid f(\cdot)$ is absolutely continuous with derivative $f^1(\cdot) \varepsilon H$ also] ,

and

$$Af = f^1 .$$

Let B denote the operator mapping the Euclidean space in which the controls range, into H by:

$$B\, u(t) \sim B(\zeta)u(t) \ , \quad 0 < \zeta < \infty$$

and similarly

$$F\omega(t) \sim F(\zeta)\omega(t) \qquad 0 < \zeta < \infty$$

Assume now that $F(t)$ and $B(t)$ are 'locally' continuous, in $0 \leq t < \infty$. Then we claim that (1.1) is representable as (a partial differential equation!)

$$\left.\begin{aligned} \dot{x}(t) &= A\, x(t) + Bu(t) + F\omega(t) \ ; \ x(0) = 0 \ . \\ y(t) &= C\, x(t) + G\omega(t) \end{aligned}\right\} \tag{1.5}$$

(or appropriate 'Wiener-process' version), where C is the operator defined by:

$$\text{Domain of } C = [f\varepsilon H \mid f(t) \text{ is continuous in } 0 \leq t < \infty]$$

[or, $f(\cdot)$ is 'locally' continuous] and

$$C f = f(0)$$

[value at the origin of the 'continuous function' representative of $f(\cdot)$] .

We can readily show that $x(t)$ is in the domain of C because of the assumption of local continuity. On the other hand we do not need to make the 'exponential rate of growth' assumptions as in the earlier version of the representation [3]. To see this we have only to note that (1.5) has the solution: (assuming that $u(\cdot)$ is locally square integrable):

$$x(t) = \int_0^t S(t-\sigma)Bu(\sigma)d\sigma + \int_0^t S(t-\sigma)\ F\omega(\sigma)d\sigma \tag{1.6}$$

where $S(t)$ is the semigroup generated by A . Now

$$h(t) = \int_0^t S(t-\sigma)\ Bu(\sigma)d\sigma \quad \text{is the function:}$$

$$h(t,\zeta) = \int_0^t B(\zeta+t-\sigma)\ u(\sigma)d\sigma \qquad 0 < \zeta < \infty$$

and $h(t,\zeta)$ is locally continuous in $0 \leq \zeta < \infty$, because of the local continuity of $B(\cdot)$. Hence $h(t)$ is in the domain of C , for each t . Moreover

$$C\ h(t) = \int_0^t B(t-\sigma)\ u(\sigma)d\sigma\ .$$

Similarly

$$C \int_0^t S(t-\sigma)\ F\omega(\sigma)d\sigma = \int_0^t F(t-\sigma)\omega(\sigma)d\sigma$$

which suffices to prove the representation. Of course to complete the representation we have that the cost functional (1.2) can be written:

$$E \int_0^t [QCx(t),\ QCx(t)]dt + E \int_0^t [u(t),\ u(t),\ u(t)]dt \tag{1.7}$$

In this form we have a stochastic control problem in a Hilbert space, and we may apply the techniques of [4]; except for the complication that C is now unbounded, uncloseable. The 'operators' B and F are Hilbert-Schmidt and in this sense there is a simplification.

Even though C is uncloseable, let us note that

$$Cx(t) = \int_0^t B(t-\sigma)\ u(\sigma)d\sigma + \int_0^t F(t-\sigma)\omega(\sigma)d\sigma$$

and hence is actually locally continuous in $0 \leq t$, and

$$g(\rho) = \int_0^\rho C\ S(\rho-\sigma)\quad F\mathit{f}(\sigma)d\sigma \qquad 0 < \rho < t$$

defines a linear bounded transformation on

$$W_n(t) = L_2\ ((0,t),R_n)$$

where R_n is the Euclidean space in which $\omega(t)$ ranges, into

$$W_0(t) = L_2((0,t);R_n)$$

for each $0 < t$. We shall only consider $u(t)$ such that

$$u(t) = \int_0^t L\ (t,s)\ y(s)ds \qquad 0 < t < T \tag{1.8}$$

where

$$g(t) = \int_0^t L\ (t,s)\ f(s)ds \qquad 0 < t < T$$

defines a Hilbert-Schmidt operator mapping $W_0(T)$ into $W_C(T)$ where

$$W_C(T) = L_2[(0,T);\ R_p]$$

where R_p is the real Euclidean space in which $u(t)$ ranges for every t. The Hilbert-Schmidtness implies that $L(t,s)$ is Hilbert-Schmidt also, a.e., and that

$$\int_0^T \int_0^t ||L(t,s)||^2_{H\cdot S}\ dt < \infty\ .$$

It is not difficult to see that

$$u(t) = \int_0^t L(t,s)y(s)ds$$

$$x(t) = \int_0^t S(t-\sigma)\ B\ u(\sigma)d\sigma + \int_0^t S(t-\sigma)\ F\omega(\sigma)d\sigma$$

$$y(t) = C\ x(t) + G\ \omega(t)$$

defines $x(\cdot)$ uniquely, for each $\omega(\cdot)$.

2. The Filtering Problem.

Let us first consider the filtering problem for (1.1) taking $u(\cdot)$ to be identically zero. We shall see that this is an essential step in solving the control problem. Thus let, in the notation of Section 1,

$$\left.\begin{aligned} x(t,\omega) &= \int_0^t S(t-\sigma)\ F\omega(\sigma)d\sigma \\ y(t,\omega) &= Cx(t,\omega) + G\omega(t) \end{aligned}\right\} \tag{2.1}$$

As we have noted earlier, the only difference from the standard problem treated in [4] is that C is uncloseable. Nevertheless since

$$Cx(t,\omega) = \int_0^t F(t-\sigma)\ \omega(\sigma)d\sigma$$

and is continuous in t for each ω , we note that, denoting by $y_t(\omega)$ the element in $W_0(t)$ defined by

$$y(s,\omega)\ , \qquad 0 < s < t$$

we see that $y_t(\omega)$ is a weak Gaussian random variable with finite second moment in $W_0(t)$ for each t . Moreover y_t has the covariance operator:

$$I + L(t)\ L(t)^*$$

where $L(t)$ is defined by

$$L(t)f = g\ ; \quad g(\rho) = \int_0^\rho F(\rho-\sigma)\ f(\sigma)d\sigma \qquad 0 < \rho < t\ ,$$

and is linear bounded on $W_n(t)$ into $W_0(t)$; and I is the identity operator on $W_0(t)$. Let

$$\hat{x}(t,\omega) = E\,[x(t,\omega) \mid y_t(\omega)]$$

Then $\hat{x}(t,\omega)$ belongs to the domain of C for each t and each ω and further

$$C\,\hat{x}(t,\omega) = E\,[Cx(t,\omega) \mid y_t(\omega)] \tag{2.2}$$

the novelty in this relation arising from the fact that C is unbounded. This can be seen readily as follows. We note that (see [4])

$$\hat{x}(t,\omega) = E\,[x(t,\omega)\, y_t(\omega)^*]\,[I + L(t)\, L(t)^*]^{-1}\, y_t(\omega) \tag{2.3}$$

where

$$E\,[x(t,\omega)\, y_t(\omega)^*] f = \int_0^t K(t,s) f(s) ds$$

where

$$K(t,\rho) = S(t-\rho) \int_0^\rho S(\rho-\sigma)F \quad F(\rho-\sigma)^*\, d\sigma$$

and the corresponding element in H is given by

$$\int_0^t \int_0^\rho F(t-\rho+\zeta)\, F(\rho-\sigma)^* d\sigma \quad f(s) ds \;, \qquad 0 < \zeta < \infty$$

and is locally continuous in $0 \leq \zeta$, for any $f(\cdot)$ in $W_0(t)$. Hence it follows that $\hat{x}(t,\omega)$ is in the domain of C for each t and ω and further a simple verification establishes (2.2) since the right side of (2.2) is given by

$$E\,[Cx(t,\omega)\, y_t(\omega)^*] \quad [I + L(t)\, L(t)^*]^{-1}$$

and for any f in $W_0(t)$:

$$E\,[Cx(t,\omega)\, y_t(\omega)^*] f = C\, E\,[x(t,\omega)\, y_t(\omega)^*] f$$

Relation (2.2) enables us to extend the arguments in [4] to show that

$$z(t,\omega) = y(t,\omega) - C\hat{x}(t,\omega) \qquad 0 < t < T$$

is again white noise. Let $P_{\delta}(t)$ denote

$$E\,[(x(t,\omega) - \hat{x}(t,\omega))\,(x(t,\omega) - \hat{x}(t,\omega))^*]\ .$$

Then $P_{\delta}(t) = E\,[x(t,\omega)\,x(t,\omega)^*] - E\,[\hat{x}(t,\omega)\,\hat{x}(t,\omega)^*]$ and it follows that $P_{\delta}(t)$ maps into the domain of C. The covariance operator of $y(\cdot)$ as an element of $W_0(T)$ has the form

$$(I + R)$$

where R is Hilbert-Schmidt and hence the Krein factorization theorem (the Kernels being strongly continuous) as in [4] yields

$$(I+R)^{-1} = (I-L)^*\,(I-L)$$

where L is Volterra and

$$z(\cdot,\omega) = (I-L)\,y(\cdot,\omega)\ .$$

Moreover

$$(I-L)^{-1} = I + M$$

where M is Hilbert-Schmidt also. Hence we can write

$$\hat{x}(\cdot,\omega) = Tz(\cdot,\omega)$$

where

$$Tf = g\ ;\quad g(t) = \int_0^t J(t,\sigma)\,z(\sigma,\omega)d\sigma$$

and following [4] we must have that

$$J(t,\sigma) = S(t-\sigma)\,(C\,P_{\delta}(\sigma))^* \tag{2.4}$$

so that

$$P_{\delta}(t)x = \int_0^t S(\sigma)F\,F^*S(\sigma)^*x d\sigma$$

$$- \int_0^t S(t-\sigma)(C\,P_{\delta}(\sigma))^*(C\,P_{\delta}(\sigma))S^*(t-\sigma)d\sigma$$

and in turn we have that, for x and y in the domain of A^*

$$[\dot{P}_f(t)x,y] = [P_f(t)x,A^*y] + [P_f(t)y,A^*x]$$
$$+ [Fx, Fy] - [C P_f(t)x, C P_f(t)y] ; \qquad (2.5)$$

$$P_f(0) = 0 .$$

Further we have:

$$\hat{x}(t,\omega) = \int_0^t S(t-\sigma) (C P_f(\sigma))^* (y(\sigma,\omega) - C\hat{x}(\sigma,\omega))d\sigma$$

$$= - \int_0^t S(t-\sigma) (C P_f(\sigma))^* C \hat{x}(\sigma,\omega)d\sigma$$

$$+ \int_0^t S(t-\sigma) (C P_f(\sigma))^* y(\sigma,\omega)d\sigma \qquad (2.6)$$

This is an 'integral equation' that $\hat{x}(t,\omega)$ satisfies. Moreover (2.6) has a unique solution. For suppose there were two solutions $\hat{x}_1(t,\omega)$, $\hat{x}_2(t,\omega)$. The difference, say $h(t)$, (fixing the ω), would satisfy

$$h(t) = - \int_0^t S(t-\sigma) (C P_f(\sigma))^* C h(\sigma)d\sigma$$

and hence we can deduce that:

$$C h(t) = - \int_0^t C S(t-\sigma) (C P_f(\sigma))^* (C h(\sigma))d\sigma .$$

But $C h(\cdot)$ is an element of $L_2(0,T)$ and the right-side defines a Hilbert-Schmidt Volterra transformation which is then quasinilpotent. Hence $C h(\cdot)$ must be zero. Hence

$$C \hat{x}_1(t,\omega) = C \hat{x}_2(t,\omega) .$$

Hence $z(t,\omega)$ remains the same:

$$z(t,\omega) = y(t,\omega) - C \hat{x}_1(t,\omega) = y(t,\omega) - C \hat{x}_2(t,\omega) .$$

But

$$\hat{x}(t,\omega) = \int_0^t J(t,\sigma)\ z(\sigma,\omega)d\sigma$$

proving the uniqueness of solution of (2.6). We could also have deduced this from the uniqueness of the Krein factorization. We can also re-write (2.6) in the differential form in the usual sense (see [4]):

$$\dot{\hat{x}}(t,\omega) = A\hat{x}(t,\omega) + (C\ P_f(t))^*(y(t,\omega) - C\hat{x}(t,\omega))$$

$$\hat{x}(0,\omega) = 0$$

yielding thus a generalization of the Kalman filter equations. Let us note in passing here that

$$A - (C\ P_f(t))^*C$$

is closed on the domain of A and the resolvent set includes the open right half plane. It does <u>not</u> however generate a <u>contraction</u> semigroup for $t > 0$.

The proof of uniqueness of solution to (2.5) can be given by invoking the dual control problem analogous to the case where C is bounded, as in [4] but will be omitted here because of limitation of space. From this it will also follow that $[P_f(t)x,x]$ is monotone in t.

Let C_n be defined on H by:

$$C_n f = g\ ;\quad g(t) = n\int_0^{1/n} f(s)ds\ .$$

Then C_n is bounded. Hence it follows that

$$E\ (C_n\ x(t,\omega))\ (C_n\ x(t,\omega)^*)$$

$$= \int_0^t (C_n\ S(\sigma)F)\ (C_n\ S(\sigma)F)^*d\sigma\quad ,$$

and as n goes to infinite, the left side converges strongly and the right side yields

$$C\ (C\ R(t,f))^*\ ;\quad R(t,t) = E\ [x(t,\omega)\ x(t,\omega)^*]\ .$$

In a similar manner we can show that

$$E\,[(C\,\hat{x}(t,\omega)\;(C\,\hat{x}(t,\omega))^*] = C\;(C\;\hat{R}(t,t))^*\;;$$

$$E\,[\hat{x}(t,\omega)\;\hat{x}(t,\omega)^*] = \hat{R}\;(t,t)$$

$$E\,[(C\;x(t,\omega) - \hat{x}(t,\omega))\;(C\;x(t,\omega) - C\;\hat{x}(t,\omega))^*]$$

$$= C\;(C\;P_{f}(t))^*$$

We are of course most interested in the case $T \to \infty$. We have seen that $[P_f(t)\;x,x]$ is monotone. Also

$$[P_f(t)x,x] \leq [R\;(t,t)x,x] = \int_0^t [S(\sigma)F\;F^*S(\sigma)^*x,x]d\sigma\;.$$

Let us assume now that

$$\int_0^\infty ||F^*S(\sigma)^*x||^2 d\sigma = [R_\infty x,x] < \infty\;. \tag{2.7}$$

(This is clearly satisfied in our example (1.4).)
Then $P_f(t)$ also converges strongly, to P_∞ , say; further P_∞ maps into the domain of C and satisfies

$$P_\infty = R_\infty - \int_0^\infty S(\sigma)\;(C\;P_\infty)^*(C\;P_\infty)\;S(\sigma)^*d\sigma$$

and hence also the algebraic equation:

$$0 = [P_\infty x,A^*y] + [P_\infty y,A^*x] + (F^*s,F^*y] - [C\;P_\infty x,C\;P_\infty y] \tag{2.8}$$

which has a unique solution.

3. The Control Problem.

Because of space limitations, we shall have to limit the presentation to the main results, emphasing only the differences arising due to the unboundedness of C . Thus, defining as in [4, Chapter 6], and confining ourselves to controls defined by (1.7);

$$x(t,\omega) - x_u(t,\omega) = \tilde{x}(t,\omega)$$

$$C\,\tilde{x}(t,\omega) + G\omega(t) = \tilde{y}(t,\omega)$$

where

$$\dot{x}_u(t,\omega) = A\,x(t,\omega) + B\,u(t,\omega)$$

we can invoke the results of section 2 to obtain that

$$z(t,\omega) = \tilde{y}(t,\omega) - C\,\hat{\tilde{x}}(t,\omega)$$

where

$$\hat{\tilde{x}}(t,\omega) = E\,[\tilde{x}(t,\omega)\mid \tilde{y}(\rho,\omega)\ ,\quad 0 < \rho < t]$$

yields white noise. We can then also proceed as in [4] to show that we can also express any $u(t,\omega)$ satisfying (1.7), also as

$$u(t,\omega) = \int_0^t m(t,\rho)\ z(\rho,\omega)d\rho$$

where the operator is Hilbert-Schmidt. The separation theorem follows easily from this, and we can show that the optimal control is given by

$$u_0(t,\omega) = -\int_t^T (Q\ C\ S(\rho-t)B)^*\ \hat{x}(\rho,\omega)d\rho \tag{3.1}$$

where

$$\hat{x}(\rho,\omega) = \hat{\tilde{x}}(\rho,\omega) + x_u(\rho,\omega)$$

and hence as in section 2, is the unique solution of

$$\dot{\hat{x}}(\rho,\omega) = A\ \hat{x}(\rho,\omega) + B\ u_0(\rho,\omega)$$

$$+ (C\ P_\delta(\rho))^*(y(\rho,\omega) - C\ \hat{x}(\rho,\omega))$$

$$\hat{x}(0,\omega) = 0\ .$$

Further we can follow [4], making appropriate modifications of the unboundedness of C, to deduce from (3.1) that

$$u_0(t,\omega) = -\ (P_c(t)B)^*\hat{x}(t,\omega) \tag{3.2}$$

where $P_c(t)$ is the solution of

$$[P_c(t)x,y] = [P_c(t)x,Ay] + [P_c(t)Ax,y]$$

$$+ [QCx,\ QCy]$$

$$-\ [(P_c(t)B)^*x,\ (P_c(t)B)^*y]\ ;$$

$$P_c(T) = 0 \tag{3.3}$$

for x,y in the domain of A.

REFERENCES

References

1. A. V. Balakrishnan: "A Note on the Structural of Optimal Stochastic Controls", Journal of Applied Mathematics and Optimization, Vol. 1, No. 1, 1974.
2. Y. Okabe:"Stationary Gaussian Processes with Markovian Property and M. Sato's Hyperfunctions", Japanese Journal of Mathematics, Vol. 41, 1973, pp. 69-122.
3. A. V. Balakrishnan: "System Theory and Stochastic Optimization", Proceedings of the NATO Advanced Institute on Network and Signal Theory, September 1972, Peter Peregrinns Lts., London.
4. A. V. Balakrishnan: Applied Functional Analysis, Springer-Verlag, 1976.
5. A. V. Balakrishnan: "Semigroup Theory and Control Theory".

"DIFFERENTIAL DELAY EQUATIONS AS CANONICAL FORMS FOR CONTROLLED HYPERBOLIC SYSTEMS WITH APPLICATIONS TO SPECTRAL ASSIGNMENT"

David L. Russell*

1. Introduction

This article is part of a continuing program of research aimed at the development of control canonical forms for certain distributed parameter control systems. This, in turn, is part of a larger effort being undertaken by a number of research workers, to arrive at a fuller understanding of the relationships between controllability of such systems and the ability to stabilize, or otherwise modify the behavior of, these systems by means of linear state feedback. (See [9], [10], [15], [11].) The present article is largely expository and will rely on the paper [12] for certain details. Nevertheless, we do present some results which go beyond those already presented in that paper.

Let us recall the control canonical form in the context of the discrete finite dimensional control system.

$$w_{k+1} = Aw_k + gu_k, \; w \in E^n, \; u \in E^1 \; . \tag{1.1}$$

If one starts with $w_0 = 0$, the control sequence $u_0, u_1, \ldots, u_{n-1}$ produces the state

* Supported in part by the Office of Naval Research under Contract No. 041-404. Reproduction in whole or in part is permitted for any purpose of the United States Government.

$$w_n = A^{n-1}gu_0 + A^{n-2}gu_1 + \cdots + Agu_{n-2} + gu_{n-1}$$

$$= (A^{n-1}g, A^{n-2}g, \ldots, Ag, g)\begin{pmatrix} u_0 \\ u_1 \\ \vdots \\ u_{n-2} \\ u_{n-1} \end{pmatrix} \equiv U\{u\} \quad . \tag{1.2}$$

The system is controllable just in case this "control to state" map is nonsingular, i.e., just in case U is a nonsingular $n\times n$ matrix. We shall assume this to be the case.

It is possible then to use the matrix U to "carry" the system (1.1) from the space E^n of state vectors w over into the space $\tilde{E}^n$ of control sequences $\{u\}$ by means of the transformation

$$w = U\tilde{\zeta} \; . \tag{1.3}$$

The transformed system is

$$\tilde{\zeta}_{k+1} = U^{-1}AU\tilde{\zeta}_k + U^{-1}gu_k$$

$$\equiv \tilde{A}\tilde{\zeta}_k + e_n u_k \; . \tag{1.4}$$

The vector e_n is the last column of the $n\times n$ identity matrix and

$$\tilde{A} = \begin{pmatrix} a^1 & 1 & 0 & \cdots & 0 \\ a^2 & 0 & 1 & \cdots & 0 \\ \vdots & \vdots & \vdots & \ddots & \vdots \\ a^{n-1} & 0 & 0 & \cdots & 1 \\ a^n & 0 & 0 & \cdots & 0 \end{pmatrix} , \tag{1.5}$$

where the a^i are the components of the vector $U^{-1}A^n g$ or, equivalently, the unique scalars for which

$$A^n g = a^1 A^{n-1}g + a^2 A^{n-2}g + \cdots + a^{n-1}Ag + a^n g \; .$$

We refer to (1.4) as the control normal form of the system (1.1). To pass to the control canonical form one employs the "convolution type" transformation

$$\tilde{\zeta} = \begin{pmatrix} 1 & 0 & 0 & \cdots & 0 \\ -a^1 & 1 & 0 & \cdots & 0 \\ -a^2 & -a^1 & 1 & \cdots & 0 \\ \vdots & \vdots & \vdots & \ddots & \vdots \\ -a^{n-1} & -a^{n-2} & -a^{n-3} & \cdots & 1 \end{pmatrix} \zeta \equiv C\zeta \tag{1.6}$$

the result of which is to produce

$$\begin{aligned} \zeta_{k+1} &= C^{-1}\tilde{A}\, C\zeta_k + C^{-1} e_n u_k \\ &\equiv \hat{A}\zeta_k + e_n u_k \end{aligned} \tag{1.7}$$

with e_n, as before, equal to $\begin{pmatrix} 0 \\ 0 \\ \vdots \\ 0 \\ 1 \end{pmatrix}$ and now

$$\hat{A} = \begin{pmatrix} 0 & 1 & 0 & \cdots & 0 \\ 0 & 0 & 1 & \cdots & 0 \\ \vdots & \vdots & \vdots & \ddots & \vdots \\ 0 & 0 & 0 & \cdots & 1 \\ a^n & a^{n-1} & a^{n-2} & \cdots & a^1 \end{pmatrix} . \tag{1.8}$$

The system (1.7) is the control canonical form for (1.1). It is significant because it enables one to see immediately the effect of linear state feedback

$$u = (k^1, k^2, \ldots, k^{n-1}, k^n) \begin{pmatrix} \zeta^1 \\ \zeta^2 \\ \vdots \\ \zeta^{n-1} \\ \zeta^n \end{pmatrix} = k\zeta$$

The closed loop system is

$$\zeta_{k+1} = (\hat{A} + e_n k)\zeta_k .$$

Since

$$\hat{A} + e_n k = \begin{pmatrix} 0 & 1 & 0 & \cdots & 0 \\ 0 & 0 & 1 & \cdots & 0 \\ \vdots & \vdots & \vdots & \ddots & \vdots \\ 0 & 0 & 0 & \cdots & 1 \\ a^n+k^1 & a^{n-1}+k^2 & a^{n-2}+k^3 & \cdots & a^1+k^n \end{pmatrix}$$

the coefficients of the characteristic polynomial of the closed loop system matrix $\hat{A} + e_n k$, and hence its eigenvalues, can be determined at will by appropriate selection of $k^1, k^2, \ldots, k^n$.

The canonical form (1.7) is equivalent to the scalar n-th order system

$$\zeta_{k+1} = a^1\zeta_k + a^2\zeta_{k-1} + \cdots + a^n\zeta_{k-n+1} + u_k .$$

In the work to follow we will see that certain infinite dimensional control systems can be reduced to a canonical form comparable to this, namely,

$$\zeta(t,2) = - e^{-\gamma}\,\zeta(t,0) + \int_0^2 \overline{p(2-\tau)\zeta(t,\tau)}\, d\tau + u(t)$$

by an entirely analogous procedure, likewise involving a "control to state" map followed by a transformation of convolution type comparable to (1.6).

2. Control Problems for Hyperbolic Systems

Let us consider the scalar hyperbolic equation

$$\frac{\partial^2 w}{\partial t^2} + \gamma \frac{\partial w}{\partial t} - \frac{\partial^2 w}{\partial x^2} + r(x)w = g(x)u(t), \qquad (2.1)$$

$$0 \le x \le 1, \quad t \ge 0$$

where γ is a constant, the real function $r \in C[0,1]$ and $g \in L^2[0,1]$. We shall suppose further that boundary conditions

$$a_0 w(0,t) + b_0 \frac{\partial w}{\partial x}(0,t) = 0, \quad a_1 w(1,t) + b_1 \frac{\partial w}{\partial x}(1,t) = 0 \tag{2.2}$$

are imposed at the endpoints $x = 0$ and $x = 1$. Throughout this article we shall suppose that $b_1 \neq 0$. Throughout the main body of the article we also assume that $b_0 = 0$, $a_0 \neq 0$, but we will comment on the case $b_0 \neq 0$ in the last section of the paper.

The Strum-Liouville operator

$$L(w) = -\frac{\partial^2 w}{\partial x^2} + r(x)w , \tag{2.3}$$

with boundary conditions of the form (2.2), $b_1 \neq 0$, $b_0 = 0$, $a_0 \neq 0$, has distinct real eigenvalues $\lambda_1 < \lambda_2 < \cdots < \lambda_k < \lambda_{k+1} < \cdots$ with (cf. [4])

$$\lambda_k = (k - \tfrac{1}{2})^2 \pi^2 + O(1), \ k \to \infty \tag{2.4}$$

and corresponding eigenfunctions ϕ_k forming an orthonormal basis for $L^2[0,1]$. Taking the inner product of (2.1) with ϕ_k we have

$$w_k''(t) + \gamma w_k'(t) + \lambda_k w_k(t) = g_k u(t), \qquad k = 1, 2, 3, \ldots , \tag{2.5}$$

where $w(x,t)$, the presumed solution of (2.1), has the expansion, convergent in $L^2[0,1]$,

$$w(x,t) = \sum_{k=1}^{\infty} w_k(t)\phi_k(x) .$$

Letting $v_k(t) = w_k'(t)$ and setting

$$\begin{pmatrix} w_k(t) \\ v_k(t) \end{pmatrix} = \begin{pmatrix} \omega_k^{-1} & \tilde{\omega}_k^{-1} \\ 1 & 1 \end{pmatrix} \begin{pmatrix} y_k(t) \\ z_k(t) \end{pmatrix} , \tag{2.6}$$

where

$$\omega_k = \frac{1}{2}\left(- \gamma + \sqrt{\gamma^2 - 4\lambda_k} \right) , \tag{2.7}$$

$$\tilde{\omega}_k = \frac{1}{2}\left(- \gamma - \sqrt{\gamma^2 - 4\lambda_k} \right) , \tag{2.8}$$

(2.5) is transformed to

$$y_k'(t) = \omega_k \, y_k(t) + h_k \, u(t) , \tag{2.9}$$

$$z_k'(t) = \tilde{\omega}_k \, z_k(t) + \tilde{h}_k \, u(t) . \tag{2.10}$$

In (2.7), (2.8) we shall use the convention that $\sqrt{\gamma^2 - 4\lambda_k}$ lies either on the non-negative real axis or the non-negative imaginary axis. The numbers h_k, $\tilde{h}_k$ in (2.9), (2.10) are

$$h_k = \frac{-\tilde{\omega}_k^{-1} g_k}{\omega_k^{-1} - \tilde{\omega}_k^{-1}} , \quad \tilde{h}_k = \frac{\omega_k^{-1} g_k}{\omega_k^{-1} - \tilde{\omega}_k^{-1}} \tag{2.11}$$

and have the property

$$\lim_{k\to\infty} \tilde{h}_k = \lim_{k\to\infty} h_k = \frac{1}{2} .$$

A slightly different transformation is used if $\gamma = 0$ and $\lambda_k = 0$ for some k (so that $\omega_k = \tilde{\omega}_k = 0$) or if $\gamma^2 = 4\lambda_k$ (so that $\omega_k = \tilde{\omega}_k$). For brevity of treatment we do not discuss these special cases here but they can be brought within the same framework.

From (2.4), (2.7) and (2.8) we see that

$$\omega_k = -\frac{\gamma}{2} + i(k - \frac{1}{2})\pi + O(\frac{1}{k}),\ k \to \infty\ , \tag{2.12}$$

$$\tilde{\omega}_k = -\frac{\gamma}{2} - i(k - \frac{1}{2})\pi + O(\frac{1}{k}),\ k \to \infty\ . \tag{2.13}$$

If we let $\omega_{-k} = \tilde{\omega}_{k+1}$, $y_{-k} = z_{k+1}$, $h_{-k} = \tilde{h}_{k+1}$, $k = 0,1,2,\ldots$ we can replace (2.9), (2.10), (2.12), (2.13) by

$$y_k' = \omega_k y_k + h_k u(t),\ -\infty < k < \infty \tag{2.14}$$

$$\omega_k = -\frac{\gamma}{2} + i(k - \frac{1}{2})\pi + O(\frac{1}{k}),\ -\infty < k < \infty\ . \tag{2.15}$$

Because the ω_k take the form (2.15) it is known (see, e.g. [6], [5], [14], [8], [13]) that the functions $e^{\omega_k t}$ form a Riesz basis (image of an orthonormal basis under a bounded and boundedly invertible linear transformation) in $L^2[0,2]$. There exists also a dual Riesz basis consisting of functions p_k, $-\infty < k < \infty$, for which

$$\int_0^2 e^{\omega_k t}\, \overline{p_\ell(2-t)}\, dt = (e^{\omega_k(\cdot)}, p_\ell)_{L^2[0,2]} = \delta_\ell^k = \begin{cases} 1, & k=\ell \\ 0, & k\neq\ell \end{cases}. \tag{2.16}$$

The biorthogonality property (2.16) enables us to study the controllability of the system (2.14) (equivalently (2.1), (2.2)) quite readily. An arbitrary control $u \in L^2[0,2]$ has the expansion

$$u(t) = \sum_{k=-\infty}^{\infty} \mu_k\, \overline{p_k(t)},\quad \sum_{k=-\infty}^{\infty} |\mu_k|^2 < \infty\ ,$$
$$\mu_k = \int_0^2 e^{\omega_k(2-t)} u(t)\, dt,\ -\infty < k < \infty\ . \tag{2.17}$$

If we begin with

$$y_k(0) = 0,\ -\infty < k < \infty$$

and apply the control $u(t)$ in (2.14), the variation of parameters formula gives

$$y_k(2) = h_k \int_0^2 e^{\omega_k(2-t)} u(t)\, dt = h_k \mu_k . \tag{2.18}$$

The first equality in (2.18) defines the control to state map for this system, i.e. the map $U : L^2[0,2] \to \ell^2$ described by

$$Uu = \left\{ h_k \int_0^2 \hat{e}^{\omega_k(2-t)} u(t)\, dt \mid -\infty < k < \infty \right\}. \tag{2.19}$$

It should be compared with the analogous matrix U in Section 1 (cf. (1.2)). We want U to be one to one, hence invertible on its range, which we accomplish with the

Approximate Controllability Assumption $h_k \neq 0,\ -\infty < k < \infty$.

The states reachable at time $t = 2$ consist of sequences

$$\{h_k \mu_k \mid -\infty < k < \infty\}, \quad \sum_{k=-\infty}^{\infty} |\mu_k|^2 < \infty ,$$

a dense subspace, which we denote by R , of ℓ^2 . In terms of the original system (2.1), (2.2) this means that we can reach states

$$w(\cdot,2) = \sum_{k=1}^{\infty} w_k \phi_k ,$$

$$v(\cdot,2) = \frac{\partial w}{\partial t}(\cdot,2) = \sum_{k=1}^{\infty} v_k \phi_k$$

with the w_k, v_k of the form

$$w_k = \frac{g_k}{k} \hat{w}_k, \quad \sum_{k=1}^{\infty} |\hat{w}_k|^2 < \infty$$

$$v_k = g_k \hat{v}_k, \quad \sum_{k=1}^{\infty} |\hat{v}_k|^2 \leq \infty .$$

We remark that it is known from [13] that the time interval of length 2 is minimal in order that U should have dense range and that the range is not altered if we take an interval of length greater than 2. Hence the choice of the interval [0,2] and the control space $L^2[0,2]$.

Our plan now is to proceed just as in Section 1. For $u \in L^2_{loc}$ the solution of (2.14) with initial state in R, as defined above, always lies in R. But $U : L^2[0,2] \xrightarrow[\text{onto}]{} R$ and U^{-1} is defined on R with $U^{-1} : R \subseteq \ell^2 \xrightarrow[\text{onto}]{} L^2[0,2]$. Hence this map can be used to transform (2.14) from a system in $R \subseteq \ell^2$ to a system in $L^2[0,2]$. In order to carry this out successfully, however, we need a certain property of the functions $e^{\omega_k t}$, $-\infty < k < \infty$, where the ω_k have the asymptotic property (2.15). We have

$$\begin{aligned} e^{\omega_k 2} + e^{-\gamma} &= e^{[-\gamma + i(2k-1)\pi + 0(\frac{1}{k})]} + e^{-\gamma} \\ &= e^{-\gamma}\left[e^{-i\pi + 0(\frac{1}{k})} + 1\right] = \varepsilon_k, \quad -\infty < k < \infty, \end{aligned} \tag{2.20}$$

where clearly $\varepsilon_k = 0(\frac{1}{k})$ and hence $\sum_{k=-\infty}^{\infty} |\varepsilon_k|^2 < \infty$.

If we now define

$$\overline{p(2-t)} = \sum_{k=-\infty}^{\infty} \varepsilon_k \overline{p_k(2-t)} \tag{2.21}$$

where the p_k are the "biorthogonal functions" defined in (2.16), we clearly have

$$e^{\omega_k 2} + e^{-\gamma} = \int_0^2 e^{\omega_k \tau} \overline{p(2-\tau)}\, d\tau, \quad -\infty < k < \infty. \tag{2.22}$$

Multiplying by $e^{\omega_k t}$ we have

$$e^{\omega_k (t+2)} + e^{-\gamma} e^{\omega_k t} = \int_0^2 e^{\omega_k (t+\tau)} \overline{p(2-\tau)}\, d\tau, \quad -\infty < k < \infty. \tag{2.23}$$

This functional differential equation is then satisfied by any linear combination $\sum_{k=-\infty}^{\infty} C_k e^{\omega_k t}$. It serves as the analog of the homogeneous version of (1.7).

The first step in the derivation of the control normal form for (2.14), which is analogous to (1.7), is to use the map U defined in (2.19), to effect the change of variable

$$y_k(t) = h_k \int_0^2 e^{\omega_k(2-\tau)} \tilde{\zeta}(t,\tau)\, d\tau = U\tilde{\zeta}(t,\cdot), \tag{2.24}$$

$$t \geq 0, \ -\infty < k < \infty .$$

The inverse map is clearly

$$\tilde{\zeta}(t,\cdot) = \sum_{k=-\infty}^{\infty} \frac{y_k(t)}{h_k} \overline{p_k(\cdot)} = U^{-1}\{y_k(t)\} \quad . \tag{2.25}$$

A formal derivation of the functional equation satisfied by $\tilde{\zeta}$ proceeds as follows.

$$\frac{\partial \tilde{\zeta}(t,\tau)}{\partial t} = \sum_{k=-\infty}^{\infty} \left(\frac{\frac{dy_k(t)}{dt}}{h_k} \right) \overline{p_k(\tau)}$$

$$= \sum_{k=-\infty}^{\infty} \left(\frac{\omega_k y_k(t) + h_k u(t)}{h_k} \right) \overline{p_k(\tau)}$$

$$= \text{(using (2.24))} \sum_{k=-\infty}^{\infty} \left(\omega_k \int_0^2 e^{\omega_k(2-s)} \tilde{\zeta}(t,s)\,ds \right) \overline{p_k(\tau)}$$

$$+ u(t) \sum_{k=-\infty}^{\infty} \overline{p_k(\tau)} .$$

Integrating by parts we now have

$$\frac{\partial\tilde{\zeta}(t,\tau)}{\partial t} = \sum_{k=-\infty}^{\infty}\left(\int_0^2 e^{\omega_k(2-s)}\,\frac{\partial\tilde{\zeta}(t,s)}{\partial s}\,ds\right)\overline{p_k(\tau)}$$

$$+ \sum_{k=-\infty}^{\infty}\left(-\tilde{\zeta}(t,2)+e^{\omega_k 2}\,\tilde{\zeta}(t,0)+u(t)\right)\overline{p_k(\tau)}\ .$$

Now using (2.22) we obtain

$$\frac{\partial\tilde{\zeta}(t,\tau)}{\partial t} = \sum_{k=-\infty}^{\infty}\left(\int_0^2 e^{\omega_k(2-s)}\left(\frac{\partial\tilde{\zeta}(t,s)}{\partial s} + \overline{p(s)}\,\tilde{\zeta}(t,0)\right)ds\right)\overline{p_k(\tau)}$$

$$+ \sum_{k=-\infty}^{\infty}\left(-\tilde{\zeta}(t,2)-e^{-\gamma}\,\tilde{\zeta}(t,0)+u(t)\right)\overline{p_k(\tau)}\ . \qquad (2.26)$$

Since the sequences $\{e^{\omega_k(2-\tau)}\}$ and $\{p_k(\tau)\}$ are dual to each other in $L^2[0,2]$, the first sum (at least formally - in general $\frac{\partial\tilde{\zeta}(t,s)}{\partial s}$ is not actually in $L^2[0,2]$) is the expansion, in terms of the functions $\overline{p_k(\tau)}$, of $\frac{\partial\tilde{\zeta}(t,\tau)}{\partial\tau} + \overline{p(\tau)}\,\tilde{\zeta}(t,0)$. The second sum can be written

$$(-\tilde{\zeta}(t,2)-e^{-\gamma}\,\tilde{\zeta}(t,0)+u(t))\sum_{k=-\infty}^{\infty}\overline{p_k(\tau)}$$

and again formally, $\sum_{k=-\infty}^{\infty}\overline{p_k(\tau)}$ can be viewed as the expansion of the distribution $\delta(\tau-2)$, since

$$\int_0^2 e^{\omega_k(2-s)}\,\delta(s-2)ds = e^{\omega_k 0} = 1,\ -\infty < k < \infty\ .$$

To avoid a multiple of $\delta(\tau-2)$ appearing on the right hand side of (2.26) we set

$$\tilde{\zeta}(t,2) + e^{-\gamma}\,\tilde{\zeta}(t,0) = u(t) \qquad (2.27)$$

and, from our earlier remarks, we now have

$$\frac{\partial\tilde{\zeta}(t,\tau)}{\partial t} = \frac{\partial\tilde{\zeta}(t,\tau)}{\partial\tau} + \overline{p(\tau)}\,\tilde{\zeta}(t,0) \ . \tag{2.28}$$

The equations (2.27), (2.28) constitute the control normal form for (2.14) (equivalently (2.1), (2.2)) and should be compared with (1.4) in Section 1.

The above formal derivation is justified in a rigorous manner in [12].

We now proceed to the control canonical form which, if (1.6) is to be paralleled, should be obtained with a "convolution type" transformation. The transformation which we use is, in fact,

$$\tilde{\zeta}(t,\cdot) = C\zeta(t,\cdot) \tag{2.29}$$

defined by

$$\tilde{\zeta}(t,\tau) = \zeta(t,\tau) - \int_0^\tau \overline{p(\tau-\sigma)}\,\zeta(t,\sigma)\,d\sigma \ . \tag{2.30}$$

With this, substitution of (2.30) into (2.27) yields

$$\zeta(t,2) = \tilde{\zeta}(t,2) + \int_0^2 \overline{p(2-\sigma)}\,\zeta(t,\sigma)\,d\sigma$$

$$= -\,e^{-\gamma}\,\tilde{\zeta}(t,0) + u(t) + \int_0^2 \overline{p(2-\sigma)}\,\zeta(t,\sigma)\,d\sigma$$

and, since (2.30) clearly gives $\tilde{\zeta}(t,0) = \zeta(t,0)$, we have

$$\zeta(t,2) + e^{-\gamma}\zeta(t,0) = u(t) + \int_0^2 \overline{p(2-\tau)}\,\zeta(t,\tau)\,d\tau \ . \tag{2.31}$$

Now substituting (2.30) into (2.28) we have, again using the fact that $\zeta(t,0) = \tilde{\zeta}(t,0)$,

$$0 = \frac{\partial}{\partial t}\left(\zeta(t,\tau) - \int_0^\tau \overline{p(\tau-\sigma)}\,\tilde{\zeta}(t,\sigma)\,d\sigma\right)$$

$$- \frac{\partial}{\partial t}\left(\zeta(t,\tau) - \int_0^\tau \overline{p(\tau-\sigma)}\,\zeta(t,\sigma)\,d\sigma\right) - \overline{p(\tau)}\,\zeta(t,0)$$

$$= \frac{\partial\zeta(t,\tau)}{\partial t} - \int_0^\tau \overline{p(\tau-\sigma)}\,\frac{\partial\zeta(t,\sigma)}{\partial t}\,d\sigma - \frac{\partial\zeta(t,\tau)}{\partial\tau}$$

$$+ \overline{p(0)}\,\zeta(t,\tau) + \int_0^\tau \frac{\partial\overline{p(\tau-\sigma)}}{\partial\tau}\,\zeta(t,\sigma)\,d\sigma - \overline{p(\tau)}\,\zeta(t,0)\ .$$

Noting that $\frac{\partial\overline{p(\tau-\sigma)}}{\partial\tau} = -\frac{\partial\overline{p(\tau-\sigma)}}{\partial\tau}$ and

integrating by parts in the second integral above we have

$$\frac{\partial\zeta(t,\tau)}{\partial t} - \frac{\partial\zeta(t,\tau)}{\partial\tau} = \int_0^\tau \overline{p(\tau-\sigma)}\left(\frac{\partial\zeta(t,\sigma)}{\partial t} - \frac{\partial\zeta(t,\sigma)}{\partial\sigma}\right)d\sigma\ .$$

This integral equation has only the solution

$$\frac{\partial\zeta(t,\tau)}{\partial t} - \frac{\partial\zeta(t,\tau)}{\partial\tau} = 0 \qquad (2.32)$$

The equations (2.31) and (2.32) together consitute the control canonical form of (2.14) (equivalently (2.1), (2.2)) and should be compared with (1.7) in Section 1. Equation (2.32) simply amounts to left translation, hence (2.31) is a neutral functional equation for ζ .

Again the above passage from the control normal form to the control canonical form has only been carried out formally, since $\overline{p(\tau-\sigma)}$ and $\zeta(t,\tau)$ do not, in general, have derivatives in $L^2[0,2]$. We again refer the reader to [12] for a more rigorous argument.

3. Spectral Determination For Hyperbolic Systems

We have noted in Section 1 that for finite dimensional systems the control canonical form is useful in establishing that the eigenvalues of the closed loop system can be placed at will with appropriate choice of the feedback row vector k . Our purpose now is to show that the canonical form developed in Section 2 can be employed to the same end with reference to the system (2.1), (2.2).

The "natural" space for study of the system (2.1) is the "finite energy" space H_E consisting of function pairs (w,v) in $H^1[0,2] \times L^2[0,2]$ with $w(0) = 0$. Supplied with an inner product

$$((w,v),(\hat{w},\hat{v}))_{H_E} = \int_0^1 [v(x)\overline{\hat{v}(x)} + w'(x)\overline{\hat{w}'(x)} + w(x)(r(x)+r_0)\overline{\hat{w}(x)}]dx, \tag{3.1}$$

(with r_0 chosen so that $r(x) + r_0 > 0$, $x \in [0,1]$) and associated norm

$$\|(w,v)\|_{H_E} = [((w,v),(w,v))_{H_E}]^{1/2} ,$$

H_E becomes a Hilbert space.

Let us consider the situation wherein the control $u(t)$ is determined by the feedback relation

$$u(t) = ((w,v),(k,\ell))_{H_E} \tag{3.2}$$

with $(k,\ell) \in H_E$. We expand w, v, k, ℓ with respect to the eigenfunctions ϕ_i of the Sturm-Liouville operator (2.3) with boundary conditions (2.2):

$$w = \sum_{i=1}^{\infty} w_i \phi_i , \quad v = \sum_{i=1}^{\infty} v_i \phi_i ,$$

$$k = \sum_{i=1}^{\infty} k_i \phi_i , \quad \ell = \sum_{i=1}^{\infty} \ell_i \phi_i ,$$

and compute, from (3.1),(3.2)

$$u(t) = \int_0^1 \left[v(x)\ \overline{\ell(x)} + \frac{\partial w}{\partial x}(x)\ \frac{\overline{\partial k}}{\partial x}(x) + w(x)(r(x)+r_0)\overline{k(x)} \right] dx$$

$$= \int_0^1 [v(x)\ \overline{\ell(x)} + (Lw+r_0 w)(x)\overline{k(x)}]\ dx$$

$$= \sum_{i=1}^{\infty} [v_i\ \overline{\ell_i} + (\lambda_i+r_0)w_i\ \overline{k_i}\ ] \ .$$

Now taking the transformation (2.6) into account we have

$$u(t) = \sum_{i=1}^{\infty} [(y_i+z_i)\overline{\ell_i}+(\lambda_i+r_0)(\omega_i^{-1}y_i+\tilde{\omega}_i^{-1}z_i)\overline{k_i}\ ]$$

$$= \sum_{i=1}^{\infty} [y_i(\overline{\ell_i}+(\lambda_i+r_0)\omega_i^{-1}\overline{k_i}\ )+z_i(\overline{\ell_i}+(\lambda_i+r_0)\tilde{\omega}_i^{-1}\overline{k_i}\ )] \qquad (3.3)$$

$$\equiv \sum_{i=1}^{\infty} [y_i\alpha_i + z_i\ \tilde{\alpha}_i] \ .$$

From the fact that $(k,\ell) \in H_E$ it can be shown quite readily (see [4], for example) that

$$\sum_{i=1}^{\infty} |\ell_i|^2 < \infty, \quad \sum_{i=1}^{\infty} \lambda_i |k_i|^2 < \infty \qquad (3.4)$$

and the conditions (3.4) are also sufficient in order that $(k,\ell) \in H_E$. From this it is easy to see that α_i and $\tilde{\alpha}_i$ in (3.3) can be chosen to be arbitrary sequences with

$$\sum_{i=1}^{\infty} |\alpha_i|^2 < \infty, \quad \sum_{i=1}^{\infty} |\tilde{\alpha}_i|^2 < \infty$$

if $(k,\ell) \in H_E$ is chosen appropriately. Thus in the context of the system (2.14) (again letting $\alpha_{-k} = \tilde{\alpha}_{k+1}$, $k = 0,1,2,\ldots$) we may assume $u(t)$ generated by the feedback law

$$u(t) = \sum_{k=-\infty}^{\infty} \alpha_k \, y_k(t) \; ,$$

with $\{\alpha_k\}$ an arbitrary sequence satisfying

$$\sum_{k=-\infty}^{\infty} |\alpha_k|^2 < \infty \; . \tag{3.5}$$

Arguing in reverse, each such $\{\alpha_k\}$ corresponds to some $(k,\ell)\in H_E$. To pass to the expression of u in terms of the variable $\tilde{\zeta}$ we use (2.24), whence

$$u(t) = \sum_{k=-\infty}^{\infty} \alpha_k \, h_k \int_0^2 e^{\omega_k(2-\tau)} \tilde{\zeta}(t,\tau)\, d\tau \; . \tag{3.6}$$

But for use in the canonical form (2.31), (2.32) we need u(t) in terms of ζ. From (2.29), (2.30) we have $\tilde{\zeta}(t,\cdot) = C\zeta(t,\cdot)$ so that

$$u(t) = \sum_{k=-\infty}^{\infty} \alpha_k h_k \int_0^2 e^{\omega_k(2-\tau)} C\zeta(t,\tau) d\tau$$

$$= \sum_{k=-\infty}^{\infty} \alpha_k h_k \int_0^2 e^{\omega_k(2-\tau)} \Big[\zeta(t,\tau) - \int_0^\tau \overline{p(2-\sigma)}\zeta(t,\sigma) d\sigma \Big]\, d\tau \; .$$

This result can be simplified as a result of the following proposition.

<u>Proposition 3.1</u> <u>There are non-zero complex numbers</u> β_k <u>with</u>

$$0 < \tilde{\beta} \le \beta_k \le \hat{\beta} \; , \quad -\infty < k < \infty \; , \tag{3.7}$$

$\tilde{\beta}$ <u>and</u> $\hat{\beta}$ <u>being independent of</u> k , <u>such that</u>

$$\int_0^2 e^{\omega_k(2-\tau)} \left[\zeta(t,\tau) - \int_0^\tau \overline{p(2-\sigma)}\zeta(t,\sigma)\, d\sigma\right] d\tau$$
$$= \beta_k \int_0^2 \overline{p_k(2-\tau)}\, \zeta(t,\tau)\, d\tau \; , \quad -\infty < k < \infty \; , \tag{3.8}$$

<u>where the functions</u> p_k <u>are the biorthogonal functions defined by</u> <u>(2.16)</u>.

Proof Since the $e^{\omega_j\tau}$ form a basis for $L^2[0,2]$ we can write

$$\zeta(t,\tau) = \sum_{j=-\infty}^{\infty} c_j e^{\omega_j\tau}, \quad c_j = (\zeta(t,\cdot), p_j)_{L^2[0,2]}$$

and it is enough to establish the result (3.8) for the special cases

$$\zeta(t,\tau) = e^{\omega_j\tau}, \quad -\infty < j < \infty .$$

Since (cf. (2.16))

$$\int_0^2 \overline{p_k(2-\tau)}\, e^{\omega_j\tau}\, d\tau = \delta_j^k$$

all we have to show is that

$$\int_0^2 e^{\omega_k(2-\tau)} \left[e^{\omega_j\tau} - \int_0^\tau \overline{p(2-\sigma)}\, e^{\omega_j\sigma}\, d\sigma \right] d\tau = \beta_j\delta_j^k . \tag{3.9}$$

For $\mathrm{Re}(s) > \mathrm{Re}(\omega_j)$ consider the expression

$$\int_0^2 e^{-s\rho} e^{\omega_j\rho}\, d\rho - \int_0^2 e^{-s\rho} \int_0^\rho \overline{p(\rho-\sigma)}\, e^{\omega_j\sigma}\, d\sigma\, d\tau$$

$$= \frac{1-e^{(-s+\omega_j)2}}{s-\omega_j} - \int_0^\infty e^{-s\rho} \int_0^\rho \overline{p(\rho-\sigma)}\, e^{\omega_j\sigma}\, d\sigma\, d\tau \tag{3.10}$$

$$+ \int_2^\infty e^{-s\rho} \int_0^\rho \overline{p(\rho-\sigma)}\, e^{\omega_j\sigma}\, d\sigma\, d\tau$$

= (letting L denote the Laplace transform, putting $\tau = r+2$ and using the fact that $\overline{p}(s) = 0$, $s > 2$)

$$= \frac{1-e^{(-s+\omega_j)2}}{s-\omega_j} - L(\bar{p})(s)\, L(e^{\omega_j\sigma})(s)$$

$$+ e^{-2s} \int_0^\infty e^{-sr} \int_r^{r+2} \overline{p(r+2-\sigma)}\, e^{\omega_j\sigma}\, d\sigma\, dr = \qquad \text{(cont.)}$$

$$= \frac{1-e^{(-s+\omega_j)2}}{s-\omega_j} + \left[e^{-2s}\int_0^2 \overline{p(2-\sigma)}e^{\omega_j\sigma}d\sigma - L(\bar{p})(s)\right]L(e^{\omega_j\sigma})(s)$$

=(using (2.22))

$$\frac{1}{s-\omega_j}\left[1-e^{(-s+\omega_j)2} + e^{-2s}\left(e^{\omega_j 2}+e^{-\gamma}\right) - L(\bar{p})(s)\right]$$

$$= \frac{1}{s-\omega_j}\left[1+e^{-2s}e^{-\gamma}-L(\bar{p})(s)\right] . \tag{3.11}$$

Since $\bar{p}$ has support in [0,2], (3.11) is entire in s, except for a (possible) pole of first order at $s = \omega_j$. For $s = \omega_k$, $k \neq j$, we have

$$\begin{aligned} &\frac{1}{\omega_k-\omega_j}\left[1+e^{-2\omega_k}e^{-\gamma}-\int_0^2 e^{-\omega_k\sigma}\bar{p}(\sigma)\,d\sigma\right] \\ &= \frac{e^{-2\omega_k}}{\omega_k-\omega_j}\left[e^{2\omega_k}+e^{-\gamma}-\int_0^2 \overline{p(2-\tau)}\,e^{\omega_k\tau}d\tau\right] = 0 , \end{aligned} \tag{3.12}$$

where we have again used (2.22). But if we take $s = \omega_k$ and multiply the right hand side of (3.10) by $e^{2\omega_k}$, (3.12) clearly shows

$$\int_0^2 e^{\omega_k(2-\tau)}\left[e^{\omega_j\tau} - \int_0^\tau \overline{p(2-\sigma)}\,e^{\omega_j\sigma}d\sigma\right]d\tau = 0, \quad k \neq j .$$

What we have shown, in effect, is that (cf. (2.29))

$$\int_0^2 e^{\omega_k(2-\tau)}\,(C(e^{\omega_j t}))(\tau)\,d\tau = 0, \quad k \neq j .$$

Since the $e^{\omega_k(2-\tau)}$, $p_j(\tau)$ form dual bases for $L^2[0,2]$,

$$C(e^{\omega_j t}) = \beta_j\,\bar{p}_j \tag{3.13}$$

for some scalar β_j and, since C is bounded and can readily be seen to be boundedly invertible, we conclude that the β_j are bounded and bounded away from zero. The result (3.8), and with it the proof of the proposition, then follows.

A more operator-theoretic proof and interpretation of Proposition 3.1 appears in [12]. One can also see from (2.22) and (3.11) that

$$\beta_j = \frac{d}{ds}\left[e^{2s}+e^{-\gamma}-e^{2s}L(\bar{p})(s)\right]\Big|_{s=\omega_j} , \qquad (3.14)$$

a result which is in agreement with the manner in which the biorthogonal functions $\bar{p}_j$ are constructed in [6], [5], [14] and which will be useful in Section 4.

Now we are in a position to prove our major result.

Theorem 3.2 Let distinct complex numbers v_j, $-\infty < \quad < \infty$, be selected with the property

$$\sum_{j=-\infty}^{\infty}\left|\frac{\omega_j - v_j}{h_j}\right|^2 < \infty . \qquad (3.15)$$

Then there is a pair $(k,\ell) \in H_E$ such that the feedback relation (3.2) yields a closed loop system

$$\frac{\partial^2\omega}{\partial t^2}+\gamma\frac{\partial w}{\partial t} - \frac{\partial^2 w}{\partial x^2} + r(x)w - g(x)\left(\left(w,\frac{\partial w}{\partial t}\right),(k,\ell)\right)_{H_E} = 0 \qquad (3.16)$$

for which the eigenvalues ω_j have been replaced by the eigenvalues v_j .

Proof Consider a system (cf. (2.14))

$$\hat{y}_j' = v_j\,\hat{y}_j, \quad -\infty < j < \infty$$

having the v_j as eigenvalues. Carrying through the transformations (2.24), (2.29) but with the ω_j replaced by the v_j , we arrive at the system (cf. (2.31), (2.32))

$$\frac{\partial\hat{\zeta}(t,\tau)}{\partial t} - \frac{\partial\hat{\zeta}(t,\tau)}{\partial\tau} = 0 \tag{3.17}$$

$$\hat{\zeta}(t,2) + e^{-\gamma}\hat{\zeta}(t,0) = \int_0^2 \overline{q(2-\tau)}\,\hat{\zeta}(t,\tau)d\tau \tag{3.18}$$

wherein (cf. (2.21), (2.16))

$$\overline{q(2-\tau)} = \sum_{j=-\infty}^{\infty} (e^{\nu_j 2}+e^{-\gamma})\,\overline{q_j(2-\tau)}\ , \tag{3.19}$$

$$\int_0^2 e^{\nu_i\tau}\,\overline{q_j(2-\tau)}\,d\tau = \delta_j^i\ . \tag{3.20}$$

Now the earlier work of the present section shows that if the feedback relation (3.2) is used in (2.31), (2.32) we obtain

$$\frac{\partial\zeta(t,\tau)}{\partial t} - \frac{\partial\zeta(t,\tau)}{\partial\tau} = 0 \tag{3.21}$$

$$\zeta(t,2) + e^{-\gamma}\zeta(t,0) = \int_0^2 \overline{p(2-\tau)}\,\zeta(t,\tau)d\tau + \sum_{j=-\infty}^{\infty} \alpha_j\beta_j h_j \int_0^2 \overline{p_j(2-\tau)}\,\zeta(t,\tau)d\tau\ . \tag{3.22}$$

To prove our theorem then we need to show the existence of a sequence $\{\alpha_j\}$ with

$$\sum_{j=-\infty}^{\infty} |\alpha_j|^2 < \infty \tag{3.23}$$

for which (3.21), (3.22) agrees with (3.17), (3.18), i.e.,

$$\sum_{j=-\infty}^{\infty} \alpha_j\beta_j h_j\,\overline{p_j(2-\tau)} = \overline{q(2-\tau)} - \overline{p(2-\tau)}\ , \tag{3.24}$$

so that (3.21), (3.22) will agree precisely with (3.17), (3.18). Let

$$\overline{q(2-\tau)} = \sum_{j=-\infty}^{\infty} d_j \, \overline{p_j(2-\tau)} , \tag{3.25}$$

Then by the biorthogonality relation (2.16),

$$d_j = \int_0^2 e^{\omega_j \tau} \, q(2-\tau) \, d\tau$$

$$= \int_0^2 e^{v_j \tau} \, \overline{q(2-\tau)} d\tau + \int_0^2 (e^{\omega_j \tau} - e^{v_j \tau}) \, \overline{q(2-\tau)} \, d\tau$$

$$= \text{(cf. (3.19), (3.20))} \quad (e^{v_j 2} + e^{-\gamma})$$

$$+ \int_0^2 \left(\int_{v_j}^{\omega_j} \frac{d}{d\sigma} e^{\sigma\tau} \, d\tau \right) \overline{q(2-\tau)} \, d\tau$$

$$= (e^{v_j 2} + e^{-\gamma}) + \int_0^2 \left(\int_{v_j}^{\omega_j} \tau e^{\sigma\tau} \, d\tau \right) \overline{q(2-\tau)} \, d\tau \tag{3.26}$$

Now

$$\left| \int_0^2 \left(\int_{v_j}^{\omega_j} \tau e^{\sigma\tau} d\tau \right) \overline{q(2-\tau)} d\tau \right| \le 2 \int_0^2 \left| \int_{v_j}^{\omega_j} e^{\sigma\tau} d\tau \right| \, |\overline{q(2-\tau)}| \, d\tau$$

$$\le 2\sqrt{2} \, |\omega_j - v_j| \, \sup |e^{\sigma\tau}| \, \|\bar{q}\|_{L^2[0,2]}$$

where the "sup" is taken over the straight line segment joining v_j and ω_j in the complex plane. From (2.15), (3.15) it is clear that this quantity is uniformly bounded, independent of j. Returning to (3.26) we see that we have

$$\begin{aligned} d_j &= (e^{\nu_j^2} + e^{-\gamma}) + \tilde{\gamma}_j |\omega_j - \nu_j| \\ &= (e^{\omega_j^2} + e^{-\gamma}) + \gamma_j |\omega_j - \nu_j| \end{aligned} \tag{3.27}$$

wherein the $\tilde{\gamma}_j$, γ_j are uniformly bounded complex numbers. Then (3.24) becomes (cf. (3.25), (3.27))

$$\sum_{j=-\infty}^{\infty} \gamma_j |\omega_j - \nu_j| \overline{p_j(2-\tau)} = \sum_{j=-\infty}^{\infty} \alpha_j \beta_j h_j \overline{p_j(2-\tau)} ,$$

whence

$$\alpha_j = \frac{\gamma_j |\omega_j - \nu_j|}{\beta_j h_j} , \quad -\infty < j < \infty .$$

Then from (3.15) and (3.7) we have (3.5) and the proof is complete.

In [12] we show that with boundary control, where g in (2.1) becomes 0 and the second equation in (2.2) becomes

$$a_1 w(1,t) + b_1 \frac{\partial w}{\partial x}(1,t) = u(t) \tag{3.28}$$

the condition (3.15) is replaced by

$$\sum_{j=-\infty}^{\infty} |\omega_j - \nu_j|^2 < \infty . \tag{3.29}$$

We also show there that the asymptotic relationship (2.15), i.e.,

$$\omega_k = -\frac{\gamma}{2} + i(k - \frac{1}{2})\pi + O(\frac{1}{k}) ,$$

can be replaced by

$$\omega_k = -\frac{\hat{\gamma}}{2} + i(k - \frac{1}{2})\pi + O(\frac{1}{k})$$

with $\hat{\gamma}$ an arbitrary complex number, by taking

$$u(t) = \hat{u}(t) + \tilde{u}(t) ,$$
$$\hat{u}(t) = a_2 w(1,t) + b_2 \frac{\partial w}{\partial x}(1,t) + C_2 \frac{\partial w}{\partial t}(1,t) \tag{3.30}$$
$$b_2 \neq -C_2 , \ \det \begin{pmatrix} a_1 & b_1 \\ a_2 & b_2 \end{pmatrix} \neq 0 .$$

After this "boundary feedback", the resulting system with the second equation of (2.2) replaced by

$$(a_1 - a_2) w(1,t) + (b_1 - b_2) \frac{\partial w}{\partial x}(1,t) - C_2 \frac{\partial w}{\partial t}(1,t) = \tilde{u}(t)$$

can be further modified by feedback similar to (3.2). The result is that a combination of feedbacks (3.30), (3.2) for boundary control (3.28) applied to (2.1) (with g=0) can produce any desired eigenvalues

$$v_k = -\frac{\hat{\gamma}}{2} + i(k - \frac{1}{2})\pi + \delta_k$$

with $\sum_{k=-\infty}^{\infty} |\delta_k|^2 < \infty$. Thus the "asymptotic line" $\mathrm{Re}(\omega) = -\frac{\gamma}{2}$ must be preserved with distributed control $g(x)u(t)$ or boundary control (3.28) with distributed feedback, but can be altered to $\mathrm{Re}(v) = -\frac{\hat{\gamma}}{2}$, $\hat{\gamma}$ arbitrary, if we allow boundary feedback as well. Then, within the established "asymptotic line" eigenvalues can be selected at will, provided the relevant condition (3.15) or (3.29) is maintained, with distributed feedback similar to (3.2). This provides a very nearly complete spectral determination theory for control systems (2.1), (2.2) (or (3.28)).

4. Spectral Determination for Certain One-Dimensional Diffusion Processes.

Let us now consider a diffusion process related to the system (2.1), namely,

$$\frac{\partial w}{\partial t} - \frac{\partial^2 w}{\partial x^2} + r(x) w = g(x) u \tag{4.1}$$

and with precisely the same boundary conditions as before, which we repeat for convenience:

$$a_0 w(0,t) + b_0 \frac{\partial w}{\partial t}(0,t) = 0, \quad a_1 w(1,t) + b_1 \frac{\partial w}{\partial t}(1,t) = 0. \tag{4.2}$$

For a system of this type it is natural to use a feedback relation of the form

$$u(t) = \int_0^1 (w'(x)\overline{k'(x)} + w(x)(r(x)+r_0)\overline{k(x)})dx \tag{4.3}$$

which corresponds to $\ell(x) \equiv 0$ in (3.2). With use of such a feedback law the closed loop system becomes

$$\frac{\partial w}{\partial t} + L_1 w = 0 , \tag{4.4}$$

where L_1 is the operator

$$(L_1 w)(x) = -\frac{\partial^2 w}{\partial x^2} + r(x)w$$

$$- g(x) \int_0^1 (w'(x)\overline{k'(x)} + w(x)(r(x)+r_0)\overline{k(x)})dx \tag{4.5}$$

with boundary conditions again of the form (4.2). Now the eigenvalues of the operator (4.5) are precisely the squares of the eigenvalues which would be obtained for the system (2.1) with $\gamma = 0$ and with the feedback control (3.2), i.e., with a special (in the context of (3.2)) feedback law for which the dependence on $v = \frac{\partial w}{\partial t}$ is zero. Let us return, therefore, to the system (2.1) with $\gamma = 0$ and explore the effect of the control law (4.3).

Since we are taking $\gamma = 0$ in (2.1) now, (2.7), (2.8) becomes

$$\omega_k = i\sqrt{\lambda_k} = -\tilde{\omega}_k = -\omega_{-k+1} , \quad k = 1, 2, 3, \ldots \tag{4.6}$$

This, together with the fact that we are taking $\ell(x) \equiv 0$, gives, in place of (3.3)

$$u(t) = \sum_{i=1}^{\infty} [y_i \alpha_i - z_i \alpha_i]$$

and then, in (2.14), we have

$$u(t) = \sum_{k=-\infty}^{\infty} \alpha_k y_k(t) , \qquad \alpha_k = - \alpha_{-k+1} , \quad k = 1,2,3,\cdots . \tag{4.7}$$

With $\gamma = 0$ we have (cf. (2.11))

$$h_k = \tilde{h}_k = g_k/2$$

and, therefore, in (2.14) we have

$$h_k = h_{-k+1} .$$

All of this means that the function $\bar{p}$ of (2.21) can be rewritten in the "symmetric form"

$$\begin{aligned}\overline{p(2-t)} &= \sum_{k=1}^{\infty} (e^{\omega_k 2}+1)\overline{p_k(2-t)} + \sum_{k=1}^{\infty} (e^{-\omega_k 2}+1)\overline{\tilde{p}_k(2-t)} \\ &= \sum_{k=1}^{\infty} (e^{\omega_k 2}+1)[\overline{p_k(2-t)} + \overline{p_k(t)}] ,\end{aligned} \tag{4.8}$$

since

$$\overline{\tilde{p}_k(2-t)} = \overline{p_{-k+1}(2-t)} = e^{2\omega_k}\, \overline{p_k(t)} .$$

Using (4.7) and (4.8) we see that the feedback relationship (4.7) now becomes

$$\begin{aligned}u(t) &= \sum_{k=1}^{\infty} \alpha_k\beta_k\, h_k\, \overline{p_k(2-t)} - \sum_{k=1}^{\infty} \alpha_k\tilde{\beta}_k\, h_k\, \overline{\tilde{p}_k(2-t)} \\ &= \sum_{k=1}^{\infty} \alpha_k\beta_k\, h_k\, \overline{p_k(2-t)} - \sum_{k=1}^{\infty} \alpha_k\tilde{\beta}_k\, h_k\, e^{2\omega_k}\, \overline{p_k(t)} .\end{aligned} \tag{4.9}$$

What we need next is a relationship between β_k and $\tilde{\beta}_k$, which we shall obtain from the formula (3.14). We have, since $\gamma = 0$,

$$e^{2s} + e^{-\gamma} - e^{2s} L(\bar{p})(s) = e^{2s} + 1 - e^{2s} L(\bar{p})(s)$$

$$= e^s [e^s + e^{-s} - e^s L(\bar{p})(s)] \ .$$

Now, since $p(t) = o$, $t > 2$,

$$e^s L(\bar{p})(s) = e^s \int_0^2 e^{-st} \overline{p(t)} dt$$

$$= (\text{setting } \tau = 2-t) = e^s \int_0^2 e^{-s(2-\tau)} \overline{p(2-\tau)} d\tau$$

$$= e^{-s} \int_0^2 e^{-(-s)\tau} \overline{p(\tau)} d\tau = e^{-s} L(\bar{p})(-s) \ .$$

Hence $e^s + e^{-s} - e^s L(\bar{p})(s)$ is an even function vanishing at $\pm\omega_1$, $\pm\omega_2$, $\pm\omega_3$, ... and from this we conclude that

$$\tilde{\beta}_k = \frac{d}{ds} \{e^s[e^s + e^{-s} - e^s L(\bar{p})(s)]\}_{s=-\omega_k}$$

$$= e^{-\omega_k} \frac{d}{ds} [e^s + e^{-s} - e^s L(\bar{p})(s)]_{s=-\omega_k}$$

$$= - e^{-\omega_k} \frac{d}{ds} [e^s + e^{-s} - e^s L(\bar{p})(s)]_{s=\omega_k}$$

$$= - e^{-2\omega_k} \frac{d}{ds} \{e^{-s}[e^s + e^{-s} - e^s L(\bar{p})(s)]\}_{s=\omega_k} = - e^{-2\omega_k} \beta_k \ .$$

Then (4.9) also assumes the symmetric form

$$u(t) = \sum_{k=1}^{\infty} \alpha_k \beta_k h_k (\overline{p_k(2-t)} + \overline{p_k(t)}) \ . \tag{4.10}$$

It is now an easy matter to see that, with this type of feedback, we can realize eigenvalues v_k for the closed loop system (4.4) provided (3.15) holds and

$$v_k = - v_{-k+1}$$

(i.e., ω_k is moved to v_k, $-\omega_k$ is moved to $-v_k$). Then, whereas the eigenvalues of L (cf. (2.3)) are $\lambda_k = \omega_k^2$, those of L_1 (cf. (4.5)) are

$$\mu_k = v_k^2, \quad k = 1, 2, 3, \ldots$$

Now $v_k = \omega_k + h_k \theta_k$, $\{\theta_k\}$ square summable, otherwise arbitrary. Hence

$$\mu_k = \lambda_k + 2\omega_k h_k \theta_k + h_k^2 \theta_k^2 = \lambda_k + \omega_k h_k \left(2\theta_k + \frac{h_k \theta_k^2}{\omega_k} \right)$$

If $\{\psi_k\}$ is a square summable sequence the equations

$$2\theta_k + \frac{h_k \theta_k^2}{\omega_k} = \psi_k, \quad k = 1, 2, 3, \ldots$$

have the square summable solution sequence

$$\theta_k = \frac{-2\omega_k + \sqrt{4\omega_k^2 + 4 h_k \omega_k \psi_k}}{2 h_k}$$

(choosing the branch of the square root which reduces to $2\omega_k$ as $h_k \to 0$). Hence we have

Theorem 4.1 <u>Let distinct complex numbers</u> μ_j, $j = 1, 2, 3, \ldots$ <u>be selected with</u>

$$\sum_{j=1}^{\infty} \left| \frac{\lambda_j - \mu_j}{\omega_j h_j} \right|^2 < \infty \, . \tag{4.11}$$

<u>Then there is a function</u> $k \in H^1[0,1]$, $k(0) = 0$, <u>such that the feedback law</u> (4.3) <u>yields a closed loop system</u> (4.4) <u>wherein the eigenvalues of</u>

<u>the operator</u> L_1 <u>are the numbers</u> μ_j, $j = 1, 2, 3, \ldots$.

Thus the λ_j can be moved "ω_j times as far" as one can move the ω_j , and "at will", subject only to (4.11).

5. Remarks on Canonical Equations of Higher Order

Almost forgotten by now is the restriction $b_0 = 0, a_0 \neq 0$, imposed at the beginning of Section 2. Nevertheless, it is a crucial restriction; for if we take $b_0 \neq 0$ (whether $a_0 \neq 0$ or not will then be immaterial) the canonical form undergoes a very decided change. To see that this is the case, consider the simple situation wherein $\gamma = 0$, $r(x) \equiv 0$, $a_0 = a_1 = 0$, $b_0 = b_1 = 1$. The eigenvalues of L (cf. (2.3)) are then

$$0,\ k^2\pi^2,\quad k = 1, 2, 3, \ldots$$

with corresponding orthonormal eigenfunctions

$$\phi_0(x) \equiv 1,\quad \phi_k(x) = \sqrt{2}\cos k\pi x,\quad k = 1, 2, 3, \ldots \ .$$

Passing to formulae analogous to (2.12), (2.13) we have

$$\omega_0 = 0,\quad \omega_k = k\pi i,$$

$$\tilde{\omega}_0 = 0,\quad \tilde{\omega}_k = -k\pi i,\quad k = 1, 2, 3, \ldots \ .$$

We have an eigenvalue of multiplicaity two at zero corresponding to the equation (cf. (2.5))

$$w_0''(t) = g_0 u(t) \tag{5.1}$$

The corresponding exponential functions (cf. (2.16) and foregoing discussion)

$$1 = e^{0t},\ e^{ik\pi t},\quad k = \pm 1, \pm 2, \ldots \tag{5.2}$$

form the familiar Fourier basis for $L^2[0,2]$. But, corresponding to the multiple eigenvalue at zero, the homogeneous counterpart of (5.1), i.e., $w_0''(t) = 0$, has independent solutions 1, t and the totality of functions which must be taken into account includes those in (5.2) and also the function t . Since those in (5.2) already form a basis for $L^2[0,2]$, we no longer have an independent set when we add the function t . The moment problem associated with the problem of controllability (cf. (2.17)) now becomes

$$y_0(2) = h_0 \int_0^2 u(t)dt \tag{5.3}$$

$$\hat{y}_0(2) = h_0 \int_0^2 (2-t)u(t)dt \tag{5.4}$$

$$y_k(2) = h_{|k|} \int_0^2 e^{ik\pi(2-t)}u(t)dt, \quad k = \pm 1, \pm 2, \ldots \tag{5.5}$$

Whereas the equations (2.17) can be solved if $y_k = h_k \mu_k$, $\sum_{k=-\infty}^{\infty} |\mu_k|^2 < \infty$, this is not possible for (5.3)-(5.5) because once $y_0(2)$, $y_k(2)$, $k = \pm 1, \pm 2, \ldots$ are selected, u is determined and thus $\hat{y}_0(2)$ is determined - it cannot be selected arbitrarily. Thus the system

$$\frac{d\hat{y}_0}{dt} = y_0 \tag{5.6}$$

$$\frac{dy_0}{dt} = h_0 u(t) \tag{5.7}$$

$$\frac{dy_k}{dt} = ik\pi y_k + h_k u(t), \quad k = \pm 1, \pm 2, \ldots \tag{5.8}$$

is not controllable using controls $u \in L^2[0,2]$. In general (see [13]) this situation can be repaired by allowing controls $u \in L^2[0,2+\varepsilon]$, $\varepsilon > 0$. But then uniqueness of control is lost. Another approach is required.

What is needed is a space of controls whose dimension is "one more than the dimension of $L^2[0,2]$" - in some appropriate sense. To make a long story (which eventually will appear elsewhere) short, the appropriate space turns out to be $H^{-1}[0,2]$, the dual space to $H^1[0,2]$ relative to $L^2[0,2]$. The control to state map, U, will carry $H^{-1}[0,2]$ into ℓ^2 and the normal form, obtained from (5.6), (5.7), (5.8) by use of U , will "live" in $H^{-1}[0,2]$. A convolution type transformation is then used to produce the control canonical form, which in this case turns out to be

$$\zeta'(t+2) = \zeta'(t) + u(t) , \tag{5.9}$$

"living" in $L^2[0,2]$ or $H^1[0,2]$, according to whether u lies in $H^{-1}[0,2]$ or $L^2[0,2]$, respectively. The "principal part" of (5.9), is obtained by simply applying to $\zeta(t+2) = \zeta(t)$, the principal part of

$$\zeta(t+2) = \zeta(t) = u(t) , \tag{5.10}$$

the operator d/dt . Of course (5.10) is the canonical form for (5.7), (5.8). The added equation (5.6) is accounted for by the differentiation in (5.9). This is entirely analogous to what happens in the finite dimensional case.

More generally, we might suppose that we have a system which, after modal analysis, takes the form

$$\frac{dz_\ell}{dt} = \mu_\ell z_\ell + f_\ell u(t), \quad \ell = 1, 2, \ldots . n \tag{5.11}$$

$$\frac{dy_k}{dt} = \omega_k y_k + h_k u(t), \quad -\infty < k < \infty \tag{5.12}$$

with the μ_ℓ , ω_k all distinct and

$$\omega_k = (k+\sigma)\pi i + O(\tfrac{1}{k}), \quad |k| \to \infty .$$

Such a system would be obtained, for example, if one considered an electronic device involving essentially no delays (such as a conventional circuit with very short wiring paths between components) attached to a long antenna in which the distributed character of the state and finite speed of signal propagation could not be ignored. If only the system (5.12) were taken into account the canonical form, obtained just as in the earlier sections of this paper, would assume the form

$$\zeta(t+2) = e^{2\sigma\pi i}\zeta(t) + \int_0^2 \overline{p(2-\tau)}\,\zeta(t+\tau)d\tau + u(t)$$

(see [12] for details). When we adjoin the system (5.11) the "natural" space of controls becomes $H^{-n}[0,2]$ and the control to state map $U : H^{-n}[0,2] \to \ell^2$. The final canonical form becomes

$$D\zeta(t+2) = e^{2\sigma\pi i}D\zeta(t) + \int_0^2 \overline{p(2-\tau)}D\zeta(t+\tau)d\tau + u(t) \qquad (5.13)$$

with D denoting the differential operator

$$\prod_{\ell=1}^{n} \left(\frac{d}{dt} - \mu_\ell\right) .$$

If some of the ω_ℓ , μ_ℓ are repeated eigenvalues, the form of some of the equations (5.11), (5.12) would have to be different if we are to have controllability but (5.13) would still be the canonical form. As in Section 3, eigenvalue assignment theorems can be obtained with the use of the form (5.13).

Finally, although we have discussed the problem of eigenvalue assignment for diffusion, or "heat", equations in Section 4, there remains the question of what the control cannonical form for such systems will eventually turn out to be. Our conjecture is that it will take the form of a "differential equation of infinite order"

$$\sum_{k=0}^{\infty} a_k \frac{d^k\zeta}{dt^k} \left(= \prod_{k=0}^{\infty} \left(1 - \frac{d/dt}{\lambda_k}\right)\zeta \right) = u(t) .$$

Comparable canonical forms may also be formed for the Euler-Bernoulli beam equations and other systems having no minimal controllability interval [2], [3], [7]).

References

1. Courant, R. and D. Hilbert: "Methods of Mathematical Physics, Vol. II - Partial Differential Equations", Interscience Pub. Co., New York, 1962.
2. Fattorini, H. O. and D. L. Russell: "Exact controllability theorems for linear parabolic equations in one space dimension", Arch. Rat. Mech. Anal., Vol. 43 (1971), pp. 272-292.
3. ________ : "Uniform bounds on biorthogonal functions for real exponentials with an application to the control theory of parabolic equations", Quart. Appl. Math., Vol. 32 (1974), pp. 45-69.
4. Graham, K. D. and D. L. Russell: "Boundary value controllability of the wave equation in a spherical region", SIAM J. Control.
5. Levinson, N.: "Gap and Density Theorems", Amer. Math. Soc. Colloq. Publ., Vol. 26 (1940), Providence, R.I.
6. Paley, R. E. A. C. and N. Wiener: "The Fourier Transform in the Complex Domain", Amer. Math. Soc. Colloq. Publ., Vol. 19 (1934), Providence, R.I.
7. Quinn, J. P.: "Time optimal control of linear distributed parameter systems", Thesis, University of Wisconsin- Madison, August 1969.
8. Riesz, F. and B. Sz.-Nagy: "Functional Analysis", F. Ungar Pub. Co., New York, 1955.
9. Russell, D. L.: "Linear stabilitzation of the linear oscillator in Hilbert space", J. Math. Anal. Appl., Vol. 25 (1969), pp. 663-675.
10. _______ : "Control theory of hyperbolic equations related to certain questions in harmonic analysis and spectral theory", Ibid., Vol. 40 (1972), pp. 336-368.
11. _______ : "Decay rates for weakly damped systems in Hilbert space obtained with control-theoretic methods", J. Diff. Eqns., Vol. 19 (1975), pp. 344-370.
12. _______ : "Canonical forms and spectral determination for a class of hyperbolic distributed parameter control systems", Technical Summary Report #1614, Mathematics Research Center, University of Wisconsin - Madison, February 1976. (Submitted to J. Math. Anal. Appl.)
13. _______ : "Nonharmonic Fourier series in the control theory of distributed parameter systems", J. Math. Anal. Appl., Vol. 18 (1967), pp, 542-559.
14. Schwartz, L.: "Étude des sommes d'exponentielles", Hermann, Paris, 1959.
15. Slemrod, M.: "A note on complete controllability and stabilizability for linear control systems in Hilbert space", SIAM J. Control, Vol. 12 (1974), pp. 500-508.

"THE TIME OPTIMAL PROBLEM FOR DISTRIBUTED CONTROL OF SYSTEMS DESCRIBED BY THE WAVE EQUATION" *

H. O. Fattorini

1. Introduction. A prototype of the problems considered here is that of stabilizing a vibrating system by means of the aplication of suitable forces during a certain time interval. To be specific, consider a uniform taut membrane clamped at the boundary Γ of a plane region Ξ. Up to the time $t = 0$ the membrane vibrates freely (no external forces are applied): thus its deflection $u(x,y,t)$ satisfies the wave equation

$$\frac{\partial^2 u}{\partial t^2} = c^2 \left(\frac{\partial^2 u}{\partial x^2} + \frac{\partial^2 u}{\partial y^2} \right), \quad ((x,y) \in \Xi\,,\ t \leq 0)$$

and the boundary condition

$$u(x,y,t) = 0 \quad ((x,y) \in \Gamma\,,\ t \leq 0)$$

where $c^2 = p/\rho$, p (resp. ρ) the modulus of elasticity (resp. the density) of the membrane.[(1)] At time $t = 0$ an external force $f(x,y,t)$ begins to be applied; the deflection of the membrane then satisfies the inhomogeneous wave equation

$$\frac{\partial^2 u}{\partial t^2} = c^2 \left(\frac{\partial^2 u}{\partial x^2} + \frac{\partial^2 u}{\partial y^2} \right) + f \quad ((x,y) \in \Xi\,,\ t \geq 0) \tag{1.1}$$

---

* This work was supported in part by the National Science Foundation under grant MPS71-02656 A04.

and the same boundary condition as for $t \le 0$,

(1.2) $u(x,y,t) = 0 \quad ((x,y) \in \Gamma,\ t \ge 0)$.

We assume that the magnitude of the force is restricted by the constraint

(1.3) $\iint_{\Xi} |f(x,y,t)|^2 \, dx\, dy \le C \quad (t \ge 0)$

(C a positive constant fixed in advance) while its objective is that of bringing the energy of the membrane

$$E(t) = \frac{1}{2} \iint_{\Xi} \left\{ \left(\frac{\partial u}{\partial t}\right)^2 + c^2 \left[\left(\frac{\partial u}{\partial x}\right)^2 + \left(\frac{\partial u}{\partial y}\right)^2 \right] \right\} dx\, dy$$

to zero in a time $T > 0$ as short as possible. In other words, we want to bring the membrane to a standstill as soon as practicable within the limitations imposed on the use of force by the constraint (1.3).

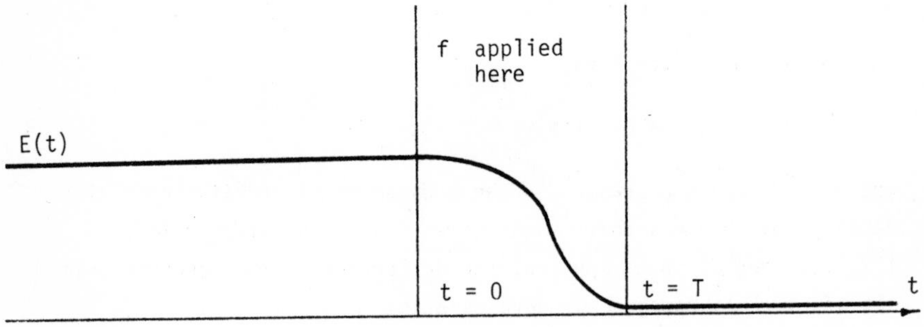

Figure 1

Three questions arise naturally in connection with this problem.

(a) Is it at all possible to reduce the energy $E(t)$ to zero in a finite time T by means of a force f subject to the constraint (1.3)?

(b) Assuming the answer to (a) is in the affirmative, does there exist a f_0 that does the transfer in minimum time?

(c) If f_0 exists, is it unique? What additional properties (say, smoothness) does it have?

Problem (a) is a typical controllability problem, and we show Section 3 that it has a solution. This is scarcely surprising in view of the extremely lavish class of controls at our disposition. (A more realistic situation would be that in which we can only use a finite number of control parameters, for instance

$$f(x,y,t) = \sum_{k=1}^{m} f_k(t)b_k(x,y)$$

where the functions $b_1, \ldots, b_m$ are fixed in advance and we can vary $f_1, \ldots, f_m$ subject to constraints of the form

$$|f_k(t)| \leq C \quad (1 \leq k \leq m, t \geq 0) .$$

This case, however, is much more complicated; the controllability problem (a) may not have a solution at all, even if we replace the final condition $E(t) = 0$ by

$$E(T) \leq \varepsilon$$

for a given $\varepsilon > 0$).

Problem (b) refers to the existence of optimal controls and it is well known that, at least in the linear case considered here, its solution follows from the solution to (a) via a simple weak compactness argument as in [2], [3], [1]. This is done in Section 4.

We examine in Section 5 problem (c). There we prove an analog of the celebrated PONTRYAGIN maximum principle in the form obtained by BELLMAN, GLICKSBERG and GROSS for linear systems in [2] and generalized to infinite-dimensional situations by BALAKRISHNAN [1] and the author [6]. The basic technique here is that of "separating hyperplanes" used in [2] for the solution of a similar control problem in finite dimensional space.

It turns out that treating the present problem directly would involve us with some of its special features (say, finite velocity of propagation of disturbances) that play no significant role on it. It is then convenient to cast it into the formalism of second order differential equations in Hilbert spaces. This is done in Section 2 and the results obtained in the following sections are then seen to be applicable to many different situations.

We examine in Section 6 some variants of the original problem obtained by replacing the constraint (1.3) on the control (which is not necessarily the only physically significant one) by other types and we show that versions of the maximum principle also hold in these cases.

2. Second-order equations in Hilbert space. We begin our quest for generality by considering the problem in Section 1 in an arbitrary number of dimensions. Let then Ξ be a bounded domain in Euclidean space $R^p (p \geq 1)$ with sufficiently smooth boundary Γ and consider the operator

$$(2.1) \qquad (Au)(x) = c^2 \Delta u(x) = c^2 \sum_{k=1}^{p} \partial_k^2 u$$

($\partial_k = \partial/\partial x_k$). The domain of A is defined in the customary way as the set of all u in the Sobolev space $H^1(\Xi)$ that satisfy the Dirichlet boundary condition

$$(2.2) \qquad u(x) = 0 \quad (x \in \Gamma)$$

and such that Δu, understood in the sense of distributions, belongs to $L^2(\Xi)$. (Recall that $H^1(\Xi))$ consists of all $u \in L^2(\Xi)$ whose distributional derivatives $\partial_1 u, \ldots, \partial_p u$ belong to $L^2(\Xi)$ with norm

$$\|u\|^2_{H^1(\Xi)} = (\|u\|^2_{L^2(\Xi)} + \sum_{k=1}^{p} \|\partial_k u\|^2_{L^2(\Xi)}) .$$

It is well known ([7]) that A is a negative definite self adjoint operator in $L^2(\Xi)$. If $u(x,t)$ is a solution of the inhomogeneous wave equation

$$\partial_t^2 u = c^2 \Delta u + f \quad (x \in \Xi, t \geq 0) \tag{2.3}$$

that satisfies the boundary condition

$$u(x,t) = 0 \quad (x \in \Gamma, t \geq 0) \tag{2.4}$$

and we denote by $u(t)$ the function in $t \geq 0$ with values in $L^2(\Xi)$ given by

$$u(t)(x) = u(x,t)$$

and define $f(t)$ similarly, then $u(\cdot)$ is (at least formally) a solution of the abstract differential equation

$$u''(t) = Au(t) + f(t) \quad (t \geq 0) . \tag{2.5}$$

We are then naturally led to the following abstract formulation of the time-optimal problem considered in Section 1: Let H be a Hilbert space and A a self adjoint operator such that

$$(Au,u) \leq - \omega\|u\|^2 \ ^{(2)} \quad (u \in D(A)) \tag{2.6}$$

for some $\omega > 0$, where $(\cdot,\cdot)$ indicates the scalar product in H. Let u_0, u_1, v_0, v_1 be given elements of H.

(a') Does there exist a control $f(\cdot)$,

$$\|f(t)\| \leq C \tag{2.7}$$

such that the corresponding solution of (2.5) with

$$u(0) = u_0 \quad u'(0) = u_1 \tag{2.8}$$

satisfies

$$u(T) = v_0 \quad u'(T) = v_1 \tag{2.9}$$

for some $T > 0$?

(b') Assuming there exists a control f satisfying the requirements in (a), does there exist a f_0 that transfers (u_0, u_1) to (v_0, v_1) in minimum time T ?

(c') What additional properties does f_0 have? (3)

In order to put the problem in a somewhat more precise footing, we must examine the equation (2.5) with some care. We start with the homogeneous equation

$$u''(t) = Au(t) \quad (t \geq 0) . \tag{2.10}$$

A <u>solution</u> of (2.10) is, by definition, a twice continuously differentiable function $u(\cdot)$ such that $u(t) \in D(A)$ for all t and (2.10) is satisfied everywhere. Solutions of (2.10) exist for "sufficiently smooth" initial data (2.8). To make this precise, define

$$C(t) = c(t,A) \quad S(t) = s(t,A) \tag{2.11}$$

where

$$c(t,\lambda) = \cos(-\lambda)^{\frac{1}{2}}t , \quad s(t,\lambda) = (-\lambda)^{\frac{1}{2}}\sin(-\lambda)^{\frac{1}{2}}t . \tag{2.12}$$ (4)

$C(t)$, $S(t)$ computed through the functional calculus for self adjoint operators ([9], Chapter XII). In view of (2.6) the spectrum of A is contained in the negative real axis, so that

$$\|C(t)\| \leq 1 , \quad \|S(t)\| \leq 1 \quad (t \geq 0) .$$

Let K be the domain of $(-A)^{\frac{1}{2}}$, the unique self adjoint, positive definite square root of $-A$. Then it is not difficult to deduce from standard functional calculus arguments that if $u_0 \in D(A)$, $u_1 \in K$,

$$u(t) = C(t)u_0 + S(t)u_1 \tag{2.13}$$

is a solution of (2.10) with initial data (2.8) and that, moreover, it is the unique such solution. As for the nonhomogeneous equation (2.5), if f is, say, continuously differentiable in $t \geq 0$ the (only) solution of (2.5) with null initial data is given by the familiar formula

$$u(t) = \int_0^t S(t-s)f(s)ds \tag{2.14}$$

(the solution with arbitrary initial data $u_0 \in D(A)$, $u_1 \in K$ is of course obtained adding (2.13) to (2.14)). However, the nature of our control problem is such that the definition of solution introduced above is too restrictive (for instance, we will be forced to consider controls f that are much less than continuously differentiable). In view of this, we proceed as follows. It is again a consequence of the functional calculus that $t \to C(t)u$ is continuous (as a H-valued function) for any $u \in H$ and continuously differentiable for $u \in K$ with $(C(t)u)' = AS(t)u$; note that $S(t)$ maps H into K (thus K into $D(A)$) and $AS(t)u$ is continuous for any $u \in K$). Also, $t \to S(t)u$ is continuously differentiable for any $u \in H$ with derivative $(S(t)u)' = C(t)u$. Making use of all these facts we extend the previous notion of solution in a way customary in control theory, namely we define

$$u(t) = C(t)u_0 + S(t)u_1 + \int_0^t S(t-s)f(s)ds \tag{2.15}$$

to be the (weak) solution of (2.5), (2.8) whenever $u_0 \in K$, $u_1 \in H$ and f is a strongly measurable, locally integrable function with values in H.[5] It is not difficult to see, on the basis of the previous observations, that $u(\cdot)$ is continuously differentiable (with derivative

$$u'(t) = AS(t)u_0 + C(t)u_1 + \int_0^t C(t-s)f(s)ds \ ) \tag{2.16}$$

and that the initial conditions (2.8) are satisfied. It is not in general true that u can be differentiated further, so that it may not be a solution of (2.5) in the original sense.

2.1 Remark. In the case where A is defined by (2.1), (2.2) the functional calculus definitions of $(-A)^{1/2}$, $C(t)$, $S(t)$ can be explicited as follows. Let $\{-\lambda_n\}$ $(0 < \lambda_0 \leq \lambda_1 \leq \dots)$ be the eigenvalues of A, $\{\varphi_n\}$ a corresponding orthonormal set of eigenfunctions. Then

$$(-A)^{\frac{1}{2}}u = \sum_{k=1}^{\infty} \lambda_n^{\frac{1}{2}}(u,\varphi_n)\varphi_n \ , \tag{2.17}$$

the domain of $(-A)^{1/2}$ consisting of all $u \in E$ such that the series on the right-hand side of (2.17) converges, or, equivalently, such that $\Sigma\lambda_n(u,\varphi_n)^2 < \infty$. We also have

$$C(t)u = \sum_{n=0}^{\infty} (\cos \lambda_n^{\frac{1}{2}}t)(u,\varphi_n)\varphi_n$$

$$S(t)u = \sum_{n=0}^{\infty} (\lambda_n^{-\frac{1}{2}}\sin \lambda_n^{\frac{1}{2}}t)(u,\varphi_n)\varphi_n$$

for all $u \in L^2(\Omega)$.

2.2 Remark. Some of the assumptions in this section (as, for example, (2.6) or the restriction of A to the class of self adjoint operators) can be weakened without modifying many of the conclusions in the next sections. We comment on this in §6.

3. Solution of the controllability problem. We look now to problem (a) in §1 in its abstract form (a'). Its solution involves finding an H-valued function $f(\cdot)$ satisfying

$$\|f(t)\| \leq C \quad (t \geq 0) \tag{3.1}$$

and such that, for some $T > 0$, the solution of (2.5) with preassigned initial data $u(0) = u_0$, $u'(0) = u_0$ satisfies $u(T) = u'(T) = 0$; in other words, such that

$$(3.2) \qquad \int_0^T S(T-t)f(t)dt = -C(T)u_0 - S(T)u_1$$

$$(3.3) \qquad \int_0^T C(T-t)f(t)dt = -AS(T)u_0 - C(T)u_1 \ .$$

Existence of a solution to (3.2), (3.3) for T large enough will follow from some simple manipulations with $C(\cdot)$ and $S(\cdot)$. We begin by introducing some useful notations. Let $\mathcal{K} = K \times H$ endowed with the norm

$$\|(u,v)\|_{\mathcal{K}}^2 = \|u\|_K^2 + \|v\|_H^2 ,$$

where the norm in K is defined by $\|u\| = \|(-A)^{1/2}u\|_H$. It is immediate that $\mathcal{K}$ is a Hilbert space. Elements of $\mathcal{K}$ will be denoted by row vectors or column vectors as convenience indicates. We denote by $\mathcal{S}(t)$ the operator from H into $\mathcal{K}$ defined by

$$\mathcal{S}(t)u = \begin{pmatrix} S(t)u \\ C(t)u \end{pmatrix}$$

and observe that, in this notation, the two equations (3.2), (3.3) can be condensed into the single equation

$$(3.4) \qquad \int_0^T \mathcal{S}(T-t)f(t)dt = -\begin{pmatrix} C(T)u_0 + S(T)u_1 \\ AS(T)u_0 + C(T)u_1 \end{pmatrix} .$$

Let now φ,ψ be twice continuously differentiable scalar functions in $0 \le t \le T$ such that

$$\varphi(0) = 0 \quad \varphi(T) = -1 \quad \varphi'(0) = 0 \quad \varphi'(T) = 0$$

$$\psi(0) = 0 \quad \psi(T) = 0 \quad \psi'(0) = 0 \quad \psi'(T) = -1 .$$

If $u \in D(A)$ both $t \to S(t)u$ and $t \to C(t)u$ are twice continuously differentiable and $S'(t)u = C(t)u$, $S''(t)u = AS(t)u$, $S(0)u = 0$, $S'(0)u = u$; $C'(t)u = AS(t)u$, $C''(t)u = AC(t)u$, $C(0)u = u$, $C'(0)u = 0$

(see the comments preceding (2.15). Then integration by parts shows that, if $u, v \in D(A)$ and

$$f(t) = \varphi(t)Au - \varphi''(t)u + \psi(t)Av - \psi''(t)v$$

we have

$$(3.5) \qquad \int_0^T S(T-t)f(t)dt = \begin{pmatrix} u \\ v \end{pmatrix}$$

and it is easy to see that we can choose φ, ψ in such a way that

$$(3.6) \qquad \|f(t)\| \le M(T)(\|u\| + \|Au\| + \|v\| + \|Av\|)$$

$M(T)$ a nonincreasing function in $T > 0$ which does not depend on u,v.

We perform now some computations with $C(\cdot)$ and $S(\cdot)$. It follows directly from its definition that $C(\cdot)$ satisfies the "cosine functional equation"

$$(3.7) \qquad C(\xi)C(\eta) = \frac{1}{2} C(\xi+\eta) + \frac{1}{2} C(\xi-\eta)$$

for all ξ, η. We obtain also from the definition of $S(\cdot)$ or by writing (3.7) for ξ', η and integrating with respect to ξ' in $0 \le \xi' \le \xi$ that

$$(3.8) \qquad S(\xi)C(\eta) = \frac{1}{2} S(\xi+\eta) + \frac{1}{2} S(\xi-\eta) .$$

We apply next both sides of (3.7) to an element of H of the form Au, $u \in D(A)$, obtaining

$$S(\xi)C(\eta)\, Au = \frac{1}{2} S''(\xi+\eta)u + \frac{1}{2} S''(\xi-\eta)u$$

$$= \frac{1}{2} C'(\xi+\eta)u + \frac{1}{2} C'(\xi-\eta)u$$

and, upon replacing η by η' and integrating in the interval $0 \le \eta' \le \eta$,

$$(3.9)\qquad S(\xi)S(\eta)Au = \frac{1}{2}\,C(\xi+\eta)u - \frac{1}{2}\,C(\xi-\eta)u \ .$$

We replace now ξ by $T - t$, η by t in (3.7), (3.8) and (3.9) and integrate in $0 \le t \le T$, obtaining

$$\int_0^T C(T-t)C(t)u\ dt = \frac{T}{2}\,C(T)u + \frac{1}{2}\,S(T)u$$

$$\int_0^T S(T-t)C(t)u\ dt = \int_0^T C(T-t)S(t)u\ dt = \frac{T}{2}\,S(T)u$$

$$\int_0^T S(T-t)S(t)Au\ dt = \frac{T}{2}\,C(T)u - \frac{1}{2}\,S(T)u$$

or, taking $u \in H$, $v \in K$,

$$(3.10)\qquad \int_0^T S(T-t)(C(t)u + AS(tv)\ dt$$

$$= \begin{pmatrix} \frac{T}{2}\,S(T)u + \frac{T}{2}\,C(T)v - \frac{1}{2}\,S(T)v \\ \\ \frac{T}{2}\,C(T)u + \frac{1}{2}\,S(T)u + \frac{T}{2}\,AS(T)v \end{pmatrix} \ .$$

We go back to the controllability problem assuming for the moment that u_0, $u_1 \in K$. We look for a solution of the problem of the form

$$f(t) = C(t)u + AS(t)v + f_1(t)$$

where $u \in H$, $v \in K$ and f_1 are as yet unspecified. In view of (3.10), equations (3.2) and (3.3) become

$$(3.11)\qquad \frac{T}{2}\,S(T)u + \frac{T}{2}\,C(T)v - \frac{1}{2}\,S(T)v + \int_0^T S(T-t)f_1(t)\ dt$$

$$= -\ C(T)u_0 - S(T)u_1$$

$$(3.12)\qquad \frac{T}{2}\,C(T)u + \frac{1}{2}\,S(T)u + \frac{T}{2}\,AS(T)v + \int_0^T C(T-t)f_1(t)\ dt$$

$$= -\ AS(T)u_0 - C(T)u_1 \ .$$

Clearly both equations will be satisfied if we take

$$(3.13) \qquad u = -\frac{2}{T}u_1 \quad v = -\frac{2}{T}u_0$$

and choose f_1 in such a way that

$$(3.14) \qquad \int_0^T S(T-t)f_1(t)\,dt = \frac{1}{T}\begin{pmatrix} -S(T)u_0 \\ S(T)u_1 \end{pmatrix}$$

which clearly can be done in view of (3.5) and comments preceding it and the fact that u_0 , u_1 belong to K . Now, it is not difficult to show that

$$(3.15) \qquad \|AS(t)u\| \leq \|(-A)^{\frac{1}{2}}u\|$$

for $u \in K$, so that

$$\|f(t)\| \leq \frac{2}{T}\|u_1\| + \frac{2}{T}\|(-A)^{\frac{1}{2}}u_0\|$$

$$+ \frac{1}{T}M(T)(\|u_0\| + \|(-A)^{\frac{1}{2}}u_0\| + \|u_1\| + \|(-A)^{\frac{1}{2}}u_1\|)$$

and it is then clear that $\|f(t)\| \leq C$ for T large enough. We have then proved

3.1 THEOREM. Let $u_0 \in K$, $u_1 \in H$. Then the controllability problem has a solution for sufficiently large T .

under the added assumption that $u_1 \in K$. We get rid of it next as follows. Let $\varepsilon = \frac{1}{3}$ and define, for every $u \in H$

$$e(u) = \{t \; ; \; \|C(t)u\| < \varepsilon\|u\| \; ; \; 0 \leq t \leq T\} \; .$$

In view of the cosine functional equation (3.7) for $\xi = \eta = t/2$ we have $2C(t/2)^2 = C(t) + I$. Accordingly, if $t \in e(u)$

$$2\|C(t/2)u\| \geq 2\|C(t/2)^2u\| \geq 1 - \|C(t)u\|$$

which makes it clear that if $t \in e(u)$ then $t/2 \notin e(u)$ or, in other words, that $e(u)$ and $\frac{1}{2}e(u)$ are disjoint. This means that the measure of $e(u)$ cannot exceed $2T/3$; hence

$$(3.16) \qquad \int_0^T \|C(t)u\|^2 \, dt \geq \frac{T}{27} \|u\|^2$$

for all $u \in E$. (This argument is due to GIUSTI.) Define now

$$N(T)u = \int_0^T C(t)^2 u \, dt .$$

Clearly $N(T)$ is a self adjoint operator and we can write (3.15) as follows:

$$(N(T)u,u) \geq \frac{T}{27} \|u\|^2$$

which shows that $N(T)$ is invertible and that

$$\|N(T)^{-1}\| \leq \frac{27}{T} .$$

We examine now (3.14) again in the light of the preceding comments on N . Write $f_1 = f_2 + f_3$ where

$$f_2(t) = \frac{1}{T} C(T-t)N(T)^{-1}S(T)u_1 .$$

Then it is clear that

$$\int_0^T C(T-s)f_2(s) \, ds = \frac{1}{T} S(T)u_1 .$$

Call now $v(T) = \int_0^T S(T-t)f_2(t) \, dt$ and, making use of the comments preceding (3.5) construct an f_3 such that

$$\int_0^T S(T-t)f_3(t) \, dt = -\frac{1}{T} \begin{pmatrix} S(T)u_0 + Tv(T) \\ 0 \end{pmatrix} .$$

To prove that this is possible, and that f_2 will have sufficiently small norm for T large enough we only have to show that $S(T)u_0$ -

$Tv(T) \in D(A)$ and that $\|A(S(T)u_0 - Tv(T))\|$ remains bounded as $T \to \infty$. The statement for $S(T)u_0$ is a direct consequence of (3.15) and preceding observations; as for $Tv(T)$ it can be easily proved with the help of (3.9). This ends the proof of Theorem 3.1 for the case $v_0 = v_1 = 0$. The general case can be easily deduced from the one just solved using the invariance of equation (2.5) with respect to time reversal. In fact, let $u_0, v_0 \in K$, $u_1, v_1 \in H$. Take T so large that there exists a solution f_1 (resp. f_2) of the controllability problem with (u_0,u_1) (resp. (v_0,v_1)) as initial data and zero final data in $0 \le t \le T$ with

$$\|f_1(t)\| \le \frac{C}{2} \quad (\text{resp. } \|f_2(t)\| \le \frac{C}{2}) \quad (0 \le t \le T) .$$

Then

$$f(t) = f_1(t) + f_2(T-t)$$

solves the general controllability problem.

4. Existence of optimal controls. Given $(u_0,u_1), (v_0,v_1) \in K$ we shall call any strongly measurable function f with values in H satisfying (3.1) and driving (u_0,u_1) to (v_0,v_1) in some time $T > 0$ an admissible control. We have established in the previous section that continuous admissible controls always exist: we show next that, giving up continuity in favor of measurability time optimal controls exist as well.

4.1 THEOREM. *Let $(u_0,u_1), (v_0,v_1) \in K$. Then there exists an optimal control f_0 driving (u_0,u_1) to (v_0,v_1) in minimum time T_0.*

The proof is an infinite dimensional analogue of that in [2]. Since the extension has already been carried out ([3], [1], [6]) in varying degrees of generality, we only sketch it here. Let T_0 be the infimum of all T for which there exists an admissible control f that drives (u_0,u_1) to (v_0,v_1) in time T. Choose now a sequence $\{f_n\}$ of admissible controls driving (u_0,u_1) to (v_0,v_1) in time T_n with

$$T_1 \ge T_2 \ge \cdots , \quad T_n \to T_0$$

and consider $\{f_n\}$ as elements of the space $L^2(0,T_1 ; H)$ (see [8], Chapter III) extending f_n to (T_n,T_1) by setting $f_n = 0$ there. Since the sequence $\{f_n\}$ is uniformly bounded in $L^2(0,T_1 ; H)$ there exists a subsequence (which we still denote $\{f_n\}$) that converges weakly to an f_0 which, as easily seen, must vanish in $t \geq T_0$ and must satisfy (3.1) almost everywhere. The fact that f_0 drives (u_0,u_1) to (v_0,v_1) follows from taking limits in the sequence of equalities

$$\int_0^{T_n} \mathcal{S}(T_n-t)f_n(t)\,dt = -\begin{pmatrix} C(T_n)u_0 + S(T_n)u_1 - v_0 \\ AS(T_n)u_0 + C(T_n)u_1 - v_1 \end{pmatrix}$$

which can be easily justified on the basis of the weak convergence of $\{f_n\}$ (see [3] for further details)).

5. <u>The maximum principle</u>. Let f_0 be a control joining two points (u_0,u_1) and (v_0,v_1) in minimum time T_0 and define $\Omega(=\Omega(T_0))$, the <u>isochronal set</u> (of f_0) to be the set of all $(u,v) \in H \times H$ of the form

$$\begin{pmatrix} u \\ v \end{pmatrix} = \int_0^{T_0} \mathcal{S}(T_0-s)f(s)\,ds \tag{5.1}$$

for some admissible control f (that is, for some strongly measurable f that satisfies

$$\|f(t)\| \leq C \quad \text{a.e. in} \quad t \geq 0) .$$

We assume in the sequel (as we plainly may) that $C = 1$. It is clear from the definition of Ω that $\Omega \subset K$. It is also immediate that Ω is convex[(6)]. Two crucial properties of the isochronal set are:

(i) The interior of Ω (in K) is non void.

(ii) $(w_0,w_1)=(v_0,v_1)-(C(t)u_0+S(t)u_1,AS(t)u_0+C(t)u_1)$ is a boundary point of Ω.

The proof of (i) follows essentially form that of Theorem 3.1. Let $(u,u') \in K$. By "running backwards" equation (2.10) we can assume that $(u,u') = (u(T_0),u'(T_0))$ for a solution $u(\cdot)$ of (2.10) with $(u(0),u'(0)) \in K$; precisely,

$$(5.2)\qquad \begin{pmatrix} u \\ u' \end{pmatrix} = \begin{pmatrix} C(T_0)u(0) + S(T_0)u'(0) \\ AS(T_0)u(0) + C(T_0)u'(0) \end{pmatrix}$$

where

$$(5.3)\qquad \begin{pmatrix} u(0) \\ u'(0) \end{pmatrix} = \begin{pmatrix} C(T_0)u - S(T_0)u' \\ -AS(T_0)u + C(T_0)u' \end{pmatrix}$$

(the justification of (5.2) and (5.3) is an easy consequence of formulas (3.7), (3.8) and (3.9)). According to Theorem 3.1 we can now find a control f such that

$$\int_0^{T_0} \mathcal{S}(T_0-t)f(t)\, dt = \begin{pmatrix} u \\ u' \end{pmatrix}$$

with

$$\|f(t)\| \le M\|(u(0), u'(0))\|_K \quad (0 \le t \le T_0) ,$$

M a constant independent of $(u(0),u'(0))$. But, on the other hand, it follows from (5.3) that

$$\|(u(0),u'(0)\|_K = \|(u,u')\|_K$$

so that if $\|(u,u')\|_K$ is sufficiently small the control f will be admissible. This shows that the origin is an interior point of Ω.

The proof of (ii) follows from (i). In fact, assume (w_0,w_1) is not a boundary point of Ω. Taking into account that the function $t \to C(t)u_0 + S(t)u_1$ is continuous in K and that $t \to AS(t)u_0 + C(t)u_1$ is continuous in H it is not difficult to deduce the existence of a $T_1 < T_0$ and a $r < 1$ such that

$$(5.4)\qquad \frac{1}{r}\left[\begin{pmatrix} v_0 \\ v_1 \end{pmatrix} - \begin{pmatrix} C(t)u_0 + S(t)u_1 \\ AS(t)u_0 + C(t)u_1 \end{pmatrix}\right] \in \Omega \qquad (T_1 \le t \le T_0) .$$

But this clearly means that

$$(5.5) \qquad \begin{pmatrix} v_0 \\ v_1 \end{pmatrix} - \begin{pmatrix} C(t)u_0 + S(t)u_1 \\ AS(t)u_0 + C(t)u_1 \end{pmatrix} = \int_0^{T_0} \mathcal{S}(T_0-s)f(s;t)ds$$

where $f(\cdot;t)$ is an admissible control with

$$(5.6) \qquad \|f(s;t)\| \le r \quad (0 \le s \le T_0) \ .$$

We observe next that

$$\lim_{t\to T_0} \int_0^{T_0-t} \mathcal{S}(T_0-s)f(s;t)dt = 0$$

in $\mathcal{K}$, so that making use of the remark at the beginning of this section we can construct a $g(s;t)$,

$$(5.7) \qquad \|g(s;t)\| \le 1 - r \quad (1 \le s \le t)$$

such that

$$(5.8) \qquad \int_0^t \mathcal{S}(t-s)g(s;t) = \int_0^{T_0-t} \mathcal{S}(T_0-s)f(s;t)dt$$

if t is sufficiently near T_0 . Hence

$$\int_0^{T_0} \mathcal{S}(T_0-s)f(s;t)ds$$

$$\int_0^t \mathcal{S}(t-s)(f(s+T_0-t;t)+g(s;t))ds$$

which shows, in view of (5.5), that (u_0,u_1) can be driven to (v_0,v_1) in time $t < T_0$ which contradicts the optimality of T_0 . This proves (ii).

We can now apply one of the standard separation theorems of functional analysis ([8], Chapter V) and deduce the existence of a nonzero continuous linear functional γ in κ such that

$$\gamma(u,u') \leq \gamma((w_0,w_1)) \tag{5.9}$$

for all (u,u') in the isochronal set Ω. It is easy to show that any linear functional in K must be of the form

$$\gamma(u,u') = (u_0^*,(-A)^{\frac{1}{2}}u) + (u_1^*,u')$$

for some $u_0^*,u_1^* \in H$. But then (5.9) can be written in the following form:

$$(u_0^*,(-A)^{\frac{1}{2}} \int_0^{T_0} S(T_0-t)f(t)dt) + (u_1^*, \int_0^{T_0} C(T_0-t)f(t)dt)$$

$$\leq (u_0^*,(-A)^{\frac{1}{2}} \int_0^{T_0} S(T_0-t)f_0(t)dt) + (u_1^*, \int_0^{T_0} C(T_0-t)f_0(t)dt)$$

for all admissible controls f. Since we can write

$$(u_0^*,(-A)^{\frac{1}{2}} \int_0^{T} S(T-t)f(t)dt) = \int_0^T ((-A)^{\frac{1}{2}} S(T-t)u_0^*,f(t))dt$$

and

$$(u_1^*, \int_0^T C(T-t)f(t)dt) = \int_0^T (C(T-t)u_1^*,f(t))dt$$

for any admissible control, we easily deduce the following

5.1 THEOREM. Let f_0 be a control driving (u_0,u_1) to (v_0,v_1) in minimum time T_0. Then there exist u_0^*, $u_1^* \in H$, $\|u_0^*\|^2 + \|u_1^*\|^2 > 0$ such that

$$(5.10)\qquad ((-A)^{\frac{1}{2}} S(T-t)u_0^* + C(T-t)u_1^*,\ f_0(t))$$

$$= \sup_{\|f\|\le 1}((-A)^{\frac{1}{2}} S(T-t)u_0^* + C(T-t)u_1^*, f)$$

<u>a.e. in</u> $0 \le t \le T$.

Clearly, (5.10) does not provide us with information on $f_0(t)$ at points where

$$\Phi(u_0^*,u_1^*;t) = (-A)^{\frac{1}{2}} S(T-t)u_0^* + C(T-t)u_1^* = 0 .$$

However, there can be only a finite number of these points in the interval $0 \le t \le T_0$. In fact, assume this is not the case, and let $\{t_n\}$ be a sequence of zeros of Φ in $0 \le t \le T_0$ converging to some t there. It is clear that $u(t) = A^{-2}\Phi(u_0^*,u_1^*,t)$ is a genuine solution of (2.10) with $u(t)=0$ and $u'(t)=\lim_{n\to\infty}(t_n-t)^{-1}(u(t_n)-u(t)) = 0$ which, by uniqueness, shows that $u(\cdot)$ (hence $\Phi(u_0^*,u_1^*,\cdot)$) is identically zero. This is absurd in view of the fact that u_0^* and u_1^* cannot vanish simultaneously.

We obtain then the following

5.2 COROLLARY. <u>If</u> f_0 <u>is an optimal control,</u>

$$(5.11)\qquad f_0(t) = \frac{\Phi(u_0^*,u_1^*,t)}{\|\Phi(u_0^*,u_1^*,t)\|} \quad (0 \le t \le T_0)$$

<u>except perhaps at a finite number of points</u> $t_0 < \cdots < t_n$ <u>where</u> $\Phi(u_0^*,u_1^*,t) = 0$.

Since $\Phi(u_0^*,u_1^*,t)$ is a continuous function, we see that f_0 must be piecewise continuous (i.e. continuous in the intervals $(0,t)$, ..., (t_n,T_0)) .

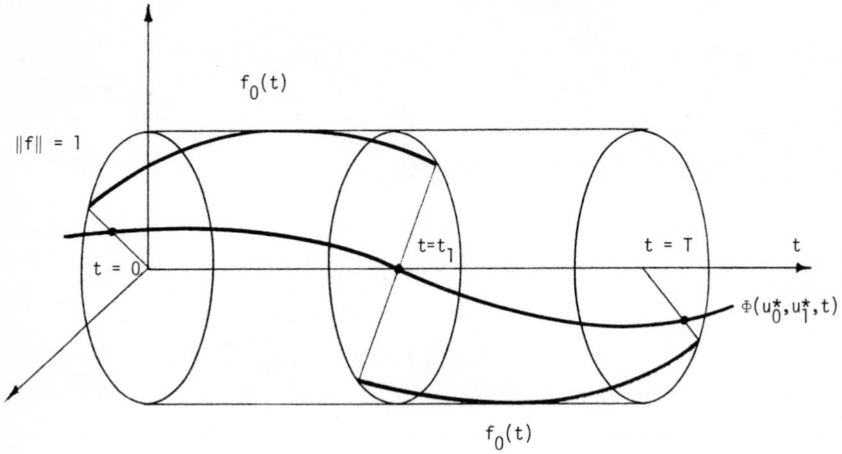

FIGURE 2

Another immediate consequence of (3.11) is <u>uniqueness</u>: in fact, let f_0, $\tilde{f}_0$ be two controls that transfer optimally (u_0,u_1) to (v_0,v_1). Then $g = 1/2(f_0+\tilde{f}_0)$ must be as well optimal and, because of (5.11) it must satisfy

$$\|g(t)\| = 1 \quad (0 \le t \le T)$$

except perhaps at a finite number of points. But then $f_0(t) = \tilde{f}_0(t)$ at every continuity point of f_0 or $\tilde{f}_0$, which shows that f_0 and $\tilde{f}_0$ must have the same discontinuity points and coincide everywhere else.

6. <u>Generalizations. The maximum principle in other geometries</u>.

We re-examine briefly the "abstract" time-optimal problem set up in Section 2. The assumptions on A there, while general enough to include the optimal problem in Section 1 are too strong to yield results in other problems. For instance, the assumption that A should satisfy (2.6) for a $\omega < 0$ does not hold for the Laplacian in the whole space R^p (in this case $\omega = 0$). This, however does not cause any significant change in the theory. In fact, we only need to assume that A is self adjoint and that (2.6) holds for some ω - perhaps negative. The operators $(-A)^{1/2}$, A^{-1}, which make several appearances in the treatment, are then replaced by $(\lambda I-A)^{1/2}$ for $\lambda > -\omega$.

A more significant generalization is that of abandoning the assumption that A is self adjoint and supposing only that the Cauchy problem for the equation (2.10) is well posed in the sense of [4], [5]. The operator-valued functions $C(\cdot)$, $S(\cdot)$ are then defined in terms of the solutions of (2.10) and it is possible to show that $(\lambda I-A)^{1/2}$, $(\lambda I-A)^{-1}$ exist for λ large enough and that all the properties needed of these entities in the treatment of the abstract time-optimal problem in Section 2 (in particular the cosine functional equation (3.7)) hold. The following caveat, however, is important: The solution of the controllability problem (a') in Section 2 that was given in Section 3 was based on uniform boundedness of $C(\cdot)$ and $S(\cdot)$ and the result does not necessarily hold without these conditions. Nevertheless, the theory can be carried forward in the following sense: _if_ the controllability problem (b') has a solution then the result in Section 4 applies to show existence of an optimal control and the results in Section 5 apply as well with some modifications. We list some of these modifications: The operator N defined in Section 3 which was used in Section 5 to show that the isochronal set Ω has interior points is now defined by

$$N(T) = \int_0^T C(t)C^*(t)u \, dt .$$

The proof that $N(T)$ has a bounded inverse runs much like in the self adjoint case, however now we take $\varepsilon = (2M(T) + 1)^{-1}$ where $M(T) = \sup\{\|C(t)\|;\ 0 \le t \le T\}$. The existence theorem 4.1 needs no modification. Finally, in Theorem 5.1 we must replace the function $\Phi(u_0^*,u_1^*,t)$ defined there by $\Phi_\lambda(u_0^*,u_1^*,t) = (\lambda I-A^*)^{1/2}S^*(T_0-t)u_0^* + C^*(T_0-t)u_1^*$, where $\lambda > -\omega$ as in the comments opening this section.

The preceding considerations are based on a desire to carry the results to what appears to be their natural level of generality; we mention here that there is no need of assuming H to be a Hilbert space(a Banach space will do) although this extension does not seem to have a wide range of applicability outside of systems described by the wave equation in one space dimension.

We show that the techniques in the present paper can be successfully applied to variants of the problem in Section 1. Assume the constraint (1.3) is replaced by

$$(6.1) \qquad \int_{\Xi\times(0,T)} |f(x,t)|^2 \, dx \, dt \le C \, .$$

The resulting time optimal problem can be cast into abstract form and treated much in the same way as the one discussed in Section 2 and 5. The solution of the controllability problem in Section 3 applies of course to the estimate (6.1), as well as the existence Theorem 4.1, with minor modifications. Call $\tilde{f}_0$ the optimal control, $\tilde{T}_0$ the driving time from $(u_0,o_1) \in K$ to $(v_0,v_1) \in K$. The isochronal set $\tilde{\Omega} = \tilde{\Omega}(\tilde{T}_0)$ is defined in the same way as Ω but using this time controls $f \in L^2(0,\tilde{T}_0;H)$ satisfying (6.1). It can be proved just as well that $\tilde{\Omega}$ is a convex subset of K with nonvoid interior and that (w_0,w_1) is a boundary point of $\tilde{\Omega}$. The separation theorem can be then applied to deduce the existence of two elements $u_0^*, u_1^* \in H$, not both zero and such that

$$(6.2) \qquad \int_0^{\tilde{T}_0} (\Phi(u_0^*,u_1^*;t),\tilde{f}_0(t))dt = \sup \int_0^{\tilde{T}_0} (\Phi(u_0^*,u_1^*;t),f(t))dt$$

where the supremum is taken with respect to all f in $L^2(0,\tilde{T}_0;H)$ with

$$\int_0^{\tilde{T}_0} \|f(t)\|^2 \, dt \le 1$$

(we are of course assuming that $C = 1$ in (6.1))and Φ denotes, as in Section 5,

$$\Phi(u_0^*,u_1^*;t) = (-A)^{\frac{1}{2}} S(\tilde{T}_0-t)u_0^* + C(\tilde{T}_0-t_0)u_1^* .$$

It turns out then that

$$(6.3) \qquad \tilde{f}_0(t) = \left(\int_0^{\tilde{T}_0} \|\Phi(u_0^*,u_1^*;t)\|^2 dt \right)^{-1/2} \Phi(u_0^*,u_1^*;t) \quad (0 \le t \le \tilde{T}_0) .$$

or we can simply write

$$(6.4) \qquad \tilde{f}_0(t) = \Phi(u_0^*,u_1^*;t) \quad (0 \le t \le \tilde{T}_0)$$

if necessary replacing u_0^*,u_1^* by $\lambda u_0^*,\lambda u_1^*$ in order that $\Phi(u_0^*,u_1^*;\cdot)$ should have norm 1 in $L^2(0,T;H)$. The representation (6.11) implies also uniqueness of the optimal control: as a matter of fact, the only information needed to prove uniqueness is that

$$(6.5) \qquad \int_0^{\tilde{T}_0} \|\tilde{f}_0(t)\|^2 \, dt = 1$$

in fact, if f_0, $\tilde{f}_0$ are two optimal controls so is $\frac{1}{2}(f_0+\tilde{f}_0) = g$: if g is to satisfy (6.5) we must have $f_0(t) = \tilde{f}_0(t)$ (note that both must be continuous in view of (6.3).

The methods outlined in the preceding remarks apply equally well to constraints of the type

$$(6.6) \qquad \int_0^T \|f(t)\|^p \, dt \le 1 ,$$

for $1 \le p \le \infty$ (although the existence theorem 4.1 has no counterpart for $p = 1$, unless H-valued measures are used as controls).

We comment finally, on two cases of some physical interest; namely, those in which the constraint (1.3) is replaced by

$$(6.7) \qquad \int_{\Xi\times(0,T)} |f(x,t)| \, dx \, dt \le C$$

or by

(6.8) $|f(x,t)| \le C$ a.e. in $\Xi\times(0,t)$.

The constraint (6.8) causes no problem as regards existence; however, the "natural" way of extending our results in Sections 3 and 5 fails, as the wave equation does not give rise to a well-posed Cauchy problem in $L^\infty(\Xi)$. The same observation applies to the constraint (6.7), in which case the "natural" setting will be the space $L^1(\Xi)$. In this case, moreover, the question of existence of optimal control seems hopeless unless the class of controls is conveniently enlarged.

REFERENCES

1. A. V. Balakrishnan, Optimal control problems in Banach spaces, SIAM Journal on Control, 3 (1965), 152-180.
2. R. Bellman, I. Glicksberg and O. Gross, On the "bang-bang" control problem, Quart. Appl. Math. 14 (1956), 11-18.
3. H. O. Fattorini, Time-optimal control of solutions of operational differential equations, SIAM Journal on Control 2 (1964), 54-59.
4. H. O. Fattorini, Ordinary differential equations in linear topological spaces, I, Jour. Diff. Equations 5 (1969), 72-105.
5. H. O. Fattorini, Ordinary differential equations in linear topological spaces, II, Jour. Diff. Equations 6 (1969), 50-70.
6. H. O. Fattorini, The time optimal control problem in Banach spaces, Appl. Math. and Optimization 1 (1974), 163-188.
7. R. Courant and D. Hilbert, Methods of Mathematical Physics, vol. 1, Interscience, N.Y., 1962.
8. N. Dunford and J. T. Schwartz, Linear Operators, Part I, Interscience, N.Y., 1957.
9. N. Dunford and J. T. Schwartz, Linear Operators, Part II, Interscience, N.Y., 1963.

FOOTNOTES

1. The past history of the membrane is obviously irrelevant and is only introduced here for the sake of picturesqueness; the relevant data are only $u(x,y,0)$, $u_t(x,y,0)$.

2. In the case at hand, (2.6) is satisfied with $\omega = \lambda_0$, $-\lambda_0 < 0$ the first eigenvalue of A . We examine later cases where (2.6) is replaced by a weaker inequality, with a view to other applications.

3. We note that the control problem (a'), (b'), (c') can be reduced to a similar control problem for a first order equation in the usual way: set $U = (u(t),u'(t))$ and write (2.5) in the form

$$U'(t) = \mathcal{A}U(t) + \mathcal{B}f(t)$$

where

$$\mathcal{A} = \begin{pmatrix} 0 & I \\ A & 0 \end{pmatrix}$$

and $\mathcal{B}$ is the operator of "projection on the second coordinate". However, the existing results on the time-optimal problem for first-order equations ([3],[6]) pertain to the case where $\mathcal{B}$ is the identity operator and thus cannot be applied here.

4. The choice of the square root is obviously irrelevant; c and s are entire functions of both their arguments.

5. The reader unfamiliar with Lebesgue-Bochner integration theory may consider f to be piecewise continuous, in which case the integral in (2.16) is an ordinary Riemann vector-valued integral. After all, it follows from the maximum principal in Section 5 that the optimal control f_0 is piecewise continuous (although the proof uses Lebesgue-Bochner integrals!)

6. Ω is also <u>bounded</u>, and an argument like the one used in the proof of Theorem 4.1 shows that it is <u>closed</u>. These properties are not significant in our analysis.

7. The somewhat overexplicit treatment of the controllability problem in Section 3 was designed in such a way that its extension to the present situation is immediate.

"SOME MAX-MIN PROBLEMS ARISING IN OPTIMAL DESIGN STUDIES"

Earl R. Barnes

1. Introduction

In 1773 Lagrange [1] attempted unsuccessfully to determine the shape of the strongest column. This is the shape of the column whose buckling load is largest among all columns of given length and volume. For columns with circular cross sections the problem was solved by Clausen [2] in 1851. It was solved for columns with general cross sections in 1962 by Tadjbakhsh and Keller [3]. Since that time similar problems have been studied by many authors. Typical among these are the problem of optimally designing vibrating beams studied in [4], and the problem of designing optimal circular arches studed in [5]. Several other interesting examples are included in [6], [7], [8] and [9]. In most of these structural design problems the critical buckling load is the lowest eigenvalue of a self-adjoint diffferential equation. The critical buckling load is therefore the minimum of a certain Rayleigh quotient. It is this minimum which has to be maximized in obtaining an optimal design. Structural design problems therefore lead to max-min problems.

Traditionally, these problems have been treated as problems in the calculus of variations. However, in many cases, it is desirable to impose constraints on the structure to be designed which make the design problem more amenable to treatment by optimal control techniques than by classical variational techniques. For example, in determining the shape of the strongest column as in [3] and [9], one is led to columns which taper to a point at various places along the column. This is clearly undesirable from a practical point of view. Therefore, in formulating the strongest column problem, a positive lower bound should be imposed on the thickness of admissible columns. The need to impose similar constraints in several other structural design problems is pointed out in [6]. Our main purpose in this paper is to provide techniques for solving optimal structural design problems in the presence of thickness constraints.

The techniques we develop have applications to a variety of other problems. They can be used to determine stability conditions for solutions of second order differential equations with periodic coefficients. Such problems arise in determining the stability or instability of an elastic structure subjected to a periodically varying force. It is a well-known fact that the stability of such a structure is dependent upon the frequency and amplitude of the applied force. We shall describe this dependence in Section 4 and give conditions on the frequency and amplitude that will guarantee stability. It turns out that the problem of determining these conditions leads to extremal eigenvalue problems. Of all Sturm-Liouville operators belonging to a certain class it is required to determine the one whose n-th eigenvalue is largest, and the one whose n-th eigenvalue is smallest, $n = 1,2,\ldots$. Since the eigenvalues are minima of Rayleigh quotients, determining the stability conditions require solving max-min and min-min problems.

A third class of max-min problems that can be solved by the method we present here arises in the design of cooling fins and spines. These are extended surfaces attached to heated bodies such as engines and radiator tubes for the purpose of dissipating heat to a surrounding medium. We shall study these problems in Section 5.

2. Extremal Eigenvalue Problems

Consider an untwisted column of length ℓ , volume V , and similar cross sections with areas $A(x)$, $0 \le x \le \ell$, satisfying

$$(2.1) \qquad a \le A(x) \le b \ , \quad 0 \le x \le \ell \ ,$$

where a and b are given positive constants. Each such column will be termed admissible. We wish to determine the strongest admissible column. This is the one whose critical buckling load is largest.

Let an admissible column be subjected to an axial load P which may cause it to buckle. In its buckled state the column will lie in a plane. Let $w(x)$ denote its lateral deflection from the straight position. Then w satisfies the equation

(2.2) $$(EI(x)w_{xx})_{xx} + Pw_{xx} = 0,\ 0 \le x \le \ell,$$

together with some set of boundary conditions such as

$$w(0) = w_x(0) = 0,$$

(2.3) $$w_{xx}(\ell) = Pw_x(\ell) + (EI(\ell)w_{xx}(\ell))_x = 0,$$

or

$$w(0) = w_x(0) = 0,$$

(2.4) $$w(\ell) = w_{xx}(\ell) = 0.$$

Conditions (2.3) correspond to a column clamped at $x=0$ and free at $x=\ell$. Conditions (2.4) correspond to a column clamped at $x=0$ and hinged at $x=\ell$.

In (2.2) E is Young's modulus of the column material and $I(x)$ is the moment of inertia of the cross section at x about a line through its centroid normal to the plane of the deflected column. Since all cross sections are similar $I(x)$ is related to $A(x)$ by

$$I(x) = \alpha A^2(x)$$

where α is a proportionality constant determined by the shape of cross sections.

Let $y(x)$, $\rho(x)$ and λ be new variables defined by $y(x) = A^2(x)w_{xx}(x)$, $\rho(x) = A^{-2}(x)$, $\lambda = \frac{P}{E\alpha}$. This change of variables transforms the differential equation (2.2) into

(2.5) $$y'' + \lambda\rho(x)y = 0\ .$$

The boundary conditions for y are obtained in [3]. They are

(2.6) $$y'(0) = 0$$
$$y(\ell) = 0$$

in the clamped-free case, and

$$(2.7)\qquad \begin{aligned} &y(0) + \ell y'(0) = 0 \\ &y(\ell) = 0 \end{aligned}$$

in the clamped-hinged case.

The column will not buckle until the load P is sufficiently great that $\lambda = \frac{P}{E\alpha}$ exceeds the lowest eigenvalue of (2.5) with appropriate boundary conditions. Let $\lambda_1(\rho)$ denote the lowest eigenvalue of (2.5) together with one set of the boundary conditions (2.6), (2.7). The optimal column design problem may be formulated as:

$$(2.8)\qquad \text{Maximize } \lambda_1(\rho)$$

subject to

$$\int_0^\ell \rho^{-1/2}(x)dx = V ,$$

$$h \le \rho(x) \le H ,$$

where $h = b^{-2}$ and $H = a^{-2}$. These constraints are the fixed volume constraint

$$(2.9)\qquad \int_0^\ell A(x)dx = V ,$$

and the limited thickness constraint

$$(2.10)\qquad a \le A(x) \le b$$

imposed on admissible columns. Problem (2.8) has been solved in [10] for the case of columns clamped at $x = 0$ and hinged at $x = \ell$.

A similar problem is solved in [9]. This is the problem of determining the optimum taper of a thin-walled tubular column of constant thickness. In this case the cross sectional areas are annuli with moment of inertia $I(x) = \alpha A^3(x)$. For columns hinged at each end the deflection is governed by the differential equation

$$(2.11)\qquad \begin{aligned} &y'' + \lambda\rho(x)y = 0 \\ &y(0) = y(\ell) = 0\ . \end{aligned}$$

The optimal design problem for such columns is

$$(2.12)\qquad \text{maximize}\quad \lambda_1(\rho)$$

subject to

$$\int_0^\ell \rho^{-1/3}(x)dx = V\ ,$$

$$h \le \rho(x) \le H\ ,$$

where $\lambda_1(\rho)$ is the lowest eigenvalue of (2.11). This problem was solved in [9] for $h = 0$ and $H = \infty$. We shall give the solution for arbitrary positive h and H in Section 3.

Problems (2.8) and (2.12) are special cases of a general class of extremal eigenvalue problems which we shall now study.

Let $\rho(x)$ be a measurable function defined on an interval $[0,\ell]$ and satisfying inequalities of the form $h \le \rho(x) \le H$, where h and H are given bounds. For each such ρ let

$$\lambda_1(\rho) < \lambda_2(\rho) < \lambda_3(\rho) < \dots$$

denote the eigenvalues of one of the boundary-value problems

$$(2.13)\qquad \begin{aligned} &y'' + \lambda\rho(x)y = 0,\ 0 \le x \le \ell, \\ &\alpha_1 y(0) + \beta_1 y'(0) = 0, \\ &\alpha_2 y(\ell) + \beta_2 y'(\ell) = 0; \end{aligned}$$

$$(2.14)\qquad \begin{aligned} &y'' + (\lambda - \rho(x))y = 0,\ 0 \le x \le \ell, \\ &\alpha_1 y(0) + \beta_1 y'(0) = 0 \\ &\alpha_2 y(\ell) + \beta_2 y'(\ell) = 0, \end{aligned}$$

where the α_i and β_i are real numbers satisfying $|\alpha_i| + |\beta_i| \neq 0$, $i = 1,2$. In the case of problem (2.13) we shall assume that $h \geq 0$. Let $f_1(x,\rho),\ldots,f_m(x,\rho)$ be m given real-valued continuous functions defined on the region $(0,\ell) \times (h,H)$, and let $c_1,\ldots,c_m$ be m given constants.

As our notation indicates, the eigenvalues $\lambda_i(\rho)$, $i = 1,2,\ldots,$ are functionals of ρ. We shall be interested in the extremal eigenvalue problems:

$$\text{(2.15)} \qquad \text{minimize } \lambda_n(\rho),\ n = 1,2,\ldots,$$

and

$$\text{(2.16)} \qquad \text{maximize } \lambda_n(\rho),\ n = 1,2,\ldots,$$

subject to

$$\text{(2.17)} \qquad \int_0^{\ell} f_i(x,\rho(x))dx = c_i,\ i = 1,\ldots,m$$

$$h \leq \rho(x) \leq H .$$

A measurable function ρ, defined on $[0,\ell]$ and satisfying these conditions will be termed admissible. The class of admissible ρ's will be denoted by A. We shall assume that A is not empty. The following theorem gives necessary conditions for an admissible ρ to be a solution of (2.15). An analogous result holds for problem (2.16).

Theorem 2.1. For a fixed n, let ρ^* be solution of (2.15), where $\lambda_n(\rho)$ refers to the n-th eigenvalue of (2.13). Let y^* denote an eigenfunction corresponding to the optimum value $\lambda_n(\rho^*)$. Then there exist Lagrange multipliers $\eta_0 \geq 0, \eta_1,\ldots,\eta_m$, not all zero, such that

$$(2.18)\qquad \max_{h \le \rho \le H} \{\eta_0 y^{*2}(x)\rho + \sum_{i=1}^{m} \eta_i f_i(x,\rho)\}$$

$$= \eta_0 y^{*2}(x)\rho^*(x) + \sum_{i=1}^{m} \eta_i f_i(x,\rho^*(x))$$

for almost all x in $[0,\ell]$.

Similarly, if $\lambda_n(\rho)$ denotes the n-th eigenvalue of (2.14), and if ρ^* is a solution of (2.15) and y^* the corresponding eigenfunction, then there exist Lagrange multipliers $\eta_0 \geq 0, \eta_1,\ldots,\eta_m$, not all zero, such that

$$(2.19)\qquad \min_{h \le \rho \le H} \{\eta_0 y^{*2}(x)\rho + \sum_{i=1}^{m} \eta_i f_i(x,\rho)\}$$

$$= \eta_0 y^{*2}(x)\rho^*(x) + \sum_{i=1}^{m} \eta_i f_i(x,\rho^*(x))$$

for almost all x in $[0,\ell]$.

The proof is based on a generalized multiplier rule found in [11, Chap. 4]. In order to state this rule we require a few definitions.

For a fixed $\rho\varepsilon A$ let

$$z_0 = \lambda_n(\rho) \quad\text{and}\quad z_i = \int_0^\ell f_i(x,\rho(x))dx, \quad i = 1,\ldots,m .$$

Let Z denote the totality of the vectors $z = (z_0,z_1,\ldots,z_m)$ obtained as ρ ranges over A . Z is a subset of Euclidean m+1-dimensional space E_{m+1} . Let z^* denote the point in Z corresponding to ρ^* .

Definition. By a derived set for Z at z^* we shall mean a set of vectors $K \subset E_{m+1}$ with the following property: If $K_1,\ldots,k_N$ is any finite collection of vectors from K , there exists a surface of the form

$$z(\varepsilon_1,\ldots,\varepsilon_N) = z^* + \sum_{j=1}^{N} k_j\varepsilon_j + o(\varepsilon),\ 0 \le \varepsilon_j \le \delta,\ j = 1,\ldots,N,$$

in Z for $\delta > 0$ and sufficiently small. $o(\varepsilon)$ is a quantity satisfying

$$\lim_{|\varepsilon|\to 0} \frac{o(\varepsilon)}{|\varepsilon|} = 0, \text{ where } |\varepsilon| = \sum_{j=1}^{N} |\varepsilon_j| .$$

Theorem 2.2. (Generalized Lagrange Multiplier Rule) Let K be a derived set for Z at z^*. Then there exist multipliers $\ell_0 \geq 0$, $\ell_1,\ldots,\ell_m$, not all zero, such that

$$\sum_{i=0}^{m} \ell_i k^i \geq 0$$

for each vector $k = (k^0,k^1,\ldots,k^m) \in K$.

This theorem is given in [11, page 178]. Its connection with Theorem 2.1 will become clear shortly. First we shall prove a lemma.

Lemma 2.1. Let ρ and σ denote any two elements in A and let $\lambda_n(\rho)$ denote the n-th eigenvalue of (2.13), $n = 1,2,\ldots$. Then

$$\lim_{t \to 0^+} \frac{\lambda_n(\rho+t(\sigma-\rho))-\lambda_n(\rho)}{t} = \frac{-\lambda_n(\rho)\int_0^\ell y^2(x)(\sigma(x)-\rho(x))dx}{\int_0^\ell \rho y^2(x)dx}$$

where y is the eigenfunction corresponding to $\lambda_n(\rho)$.

Proof. For $0 < t < 1$ let $\rho_t = \rho+t(\sigma-\rho)$ and let y_t denote the eigenfunction corresponding to $\lambda_n(\rho_t)$, normalized so that

$$\int_0^\ell \rho_t y_t^2 dx = \int_0^\ell \rho y^2 dx .$$

Then $\quad y_t'' + \lambda_n(\rho_t)\rho_t(x)y_t = 0$

and $\quad y'' + \lambda_n(\rho)\rho(x)y = 0 .$

Upon multiplying the first of these equations by y and the second by y_t and integrating by parts we obtain

$$\int_0^\ell \lambda_n(\rho_t)\rho_t(x)y_t y dx = \int_0^\ell \lambda_n(\rho)\rho(x)y y_t dx .$$

This implies that

$$\frac{\lambda_n(\rho+t(\sigma-\rho))-\lambda_n(\rho)}{t} = \frac{-\lambda_n(\rho_t)\int_0^\ell(\sigma-\rho)y_t y dx}{\int_0^\ell \rho y_t y dx} .$$

By letting $t \to 0$, and making use of the continuous dependence of the eigenvalues and eigenfunctions of (2.13) on ρ , we obtain the conclusion of the lemma.

Proof of Theorem 2.1. Since eigenfunctions are unique only up to a scalar factor we may assume that

$$\int_0^{\ell} \rho^* y^{*2}(x)dx = \lambda_n(\rho^*) .$$

Let $x \in (0,\ell)$ and $\rho \in [h,H]$ be fixed. Consider the vector $k = (k^0,k^1,\dots k^m) \in E_{m+1}$ defined by

$$(2.20) \qquad k^0 = -(\rho-\rho^*(x))y^{*2}(x) ,$$

$$k^i = f_i(x,\rho) - f_i(x,\rho^*(x)), \quad i = 1,\dots,m .$$

We shall show that, except for x's forming a set of measure zero, the vectors k form a derived set for Z at z^* .

Let $k_1,k_2,\dots,k_N$ denote an arbitrary finite collection of vectors of type (2.20). Then there exist values $x_1,x_2,\dots,x_N \in (0,\ell)$ and $\rho_1,\dots,\rho_N \in [h,H]$ such that

$$k_j^0 = -(\rho_j-\rho^*(x_j))y^{*2}(x_j)$$

and

$$k_j^i = f_i(x_j,\rho_j) - f_i(x_j,\rho^*(x_j)), \; i = 1,\dots,m ,$$

where

$$k_j = (k_j^0,k_j^1,\dots,k_j^m), \; j = 1,\dots,N .$$

For simplicity we shall assume that $x_1 \leq x_2 \leq \dots \leq x_N$ and that the x_j are points of continuity of ρ^* . The proof is valid under much weaker measurability assumptions. Let $\delta > 0$ be chosen such that $x_i + N\delta < x_j$ if $x_i < x_j$, and such that $x_N + N\delta < \ell$. Let $\varepsilon = (\varepsilon_1,\dots,\varepsilon_N)$ be a vector of real parameters satisfying $0 \leq \varepsilon_j \leq \delta$, $j = 1,\dots,N$. Let

$$X_1 = x_1, \; X_j = x_j + \varepsilon_1 + \dots + \varepsilon_{j-1} ,$$

$j = 2,\dots,N$. Clearly, the intervals $I_j = [X_j . X_j + \varepsilon_j]$ are nonoverlapping.

Define the admissible function ρ_ε by

$$\rho_\varepsilon(x) = \begin{cases} \rho^*(x), & x \notin \bigcup_{j=1}^{N} I_j \\ \rho_j, & x \in I_j, \; j = 1,\dots,N . \end{cases}$$

Let $z_0(\varepsilon) = \lambda_n(\rho_\varepsilon)$ and $z_i(\varepsilon) = \int_0^\ell f_i(x,\rho_\varepsilon(x))dx, \quad i = 1,\dots,m.$ An easy consequence of Lemma 2.1 is that

$$z_0(\varepsilon) = \lambda_n(\rho^*) - \int_0^\ell y^{*2}(x)(\rho_\varepsilon(x)-\rho^*(x))dx + o(\varepsilon)$$

$$(2.21) \qquad = \lambda_n(\rho^*) - \sum_{j=1}^{N} \varepsilon_j y^{*2}(x_j)(\rho_j - \rho^*(x_j)) + o(\varepsilon)$$

$$= z_0^* + \sum_{j=1}^{N} \varepsilon_j k_j^0 + o(\varepsilon) .$$

Here we have used the continuity property of ρ^* at the points x_j in differentiating the integral in (2.21). Similarly, we have

$$z_i(\varepsilon) = \int_0^\ell f_i(x,\rho^*(x))dx + \sum_{j=1}^{N} \varepsilon_j[f_i(x_j,\rho_j) - f_i(x_j,\rho^*(x_j))] + o(\varepsilon)$$

$$(2.22) \qquad = z_i^* + \sum_{j=1}^{N} \varepsilon_j k_j^i + o(\varepsilon), \quad i = 1,\dots,m .$$

Combining (2.21) and (2.22) we see that the vector $z(\varepsilon) = (z_0(\varepsilon), z_1(\varepsilon),\dots,z_m(\varepsilon))$ satisfies

$$z(\varepsilon) = z^* + \sum_{j=0}^{N} \varepsilon_j k_j + o(\varepsilon) \in Z .$$

This completes the proof that vectors of the form (2.20), for almost all x, form a derived set.

It now follows from Theorem 2.2 that there exist multipliers $\ell_0 \geq 0$, $\ell_1,\ldots,\ell_m$ such that

$$-\ell_0 y^{*2}(x)(\rho-\rho^*(x)) + \sum_{i=1}^{m} \ell_i[f_i(x,\rho)-f_i(x,\rho^*(x))] \geq 0$$

for all $\rho \in [h,H]$, and for almost all $x \in [0,\ell]$. By taking $\eta_0 = \ell_0$, $\eta_i = -\ell_i$, $i = 1,\ldots,m$, we obtain the conclusion of Theorem 2.1.

In a similar manner the following theorem can be proved.

Theorem 2.3. For a fixed n let ρ^* be a solution of (2.16) where $\lambda_n(\rho)$ refers to the n-th eigenvalue of (2.13). Let y^* denote the eigenfunction corresponding to the optimum value $\lambda_n(\rho^*)$. Then there exist Lagrange multipliers $\eta_0 \geq 0$, $\eta_1,\ldots,\eta_m$ such that

$$\min_{h \leq \rho \leq H} \{\eta_0 y^{*2}(x)\rho + \sum_{i=1}^{m} \eta_i f_i(x,\rho)\} = \eta_0 y^{*2}(x)\rho^*(x) + \sum_{i=1}^{m} \eta_i f_i(x,\rho^*(x)) \tag{2.23}$$

for almost all $x \in [0,\ell]$.

Similarly, if $\lambda_n(\rho)$ denotes the n-th eigenvalue of (2.14), and if ρ^* is a solution of (2.16) and y^* the corresponding eigenfunction, then there exist Lagrange multipliers $\eta_0 \geq 0$, $\eta_1,\ldots,\eta_m$ such that

$$\max_{h \leq \rho \leq H} \{\eta_0 y^{*2}(x)\rho + \sum_{i=1}^{m} \eta_i f_i(x,\rho)\} = \eta_0 y^{*2}(x)\rho^*(x) + \sum_{i=1}^{m} \eta_i f_i(x,\rho^*(x)) \tag{2.24}$$

for almost all x in $[0,\ell]$.

3. The Shape of the Strongest Tubular Column

Let ρ^* denote a solution of problem (2.12). As we have seen, ρ^* determines the shape of the strongest thin-walled tubular column in the class of columns hinged at $x = 0$ and ℓ, and having fixed length and volume, and similar cross sections. According to condition (2.23) of Theorem 2.3, there exist constants $\eta_0 \geq 0$ and η such that

$$\min_{h \leq \rho \leq H} [\eta_0 y^{*2}(x)\rho + \eta\rho^{-1/3}] = \eta_0 y^{*2}(x)\rho^*(x) + \eta\rho^{*-1/3}(x) \tag{3.1}$$

for almost all x in $[0,\ell]$.

We shall assume that the quantities a, b, V in (2.9) and (2.10) satisfy $a\ell < V < b\ell$. When this is the case it is easy to show that η_0 and η are > 0. Without loss of generality, we take $\eta_0 = 1/3$.

For convenience we shall drop the $*$ on ρ^* and y^*. Condition (3.1) implies that

$$\rho(x) = \begin{cases} h & \text{if } \eta^{3/4}(y(x))^{-3/2} \leq h \\ \eta^{3/4}(y(x))^{-3/2} & \text{if } h \leq \eta^{3/4}(y(x))^{-3/2} \leq H \\ H & \text{if } \eta^{3/4}(y(x))^{-3/2} \geq H \end{cases} \tag{3.2}$$

for almost all x in $[0,\ell]$.

Since $y(0) = 0$, for values of x sufficiently close to 0 we have $\eta^{3/4}(y(x))^{-3/2} > H$ and $\rho(x) = H$. For these values of x the differential equation (2.11) is simply

$$y'' + \lambda H y + 0 . \tag{3.3}$$

It is instructive to view the solution in the phase, or y,y' plane. In phase space, (3.3) implies that the point $(y(x),y'(x))$ is moving along the ellipse

$$y'^2 + \lambda H y^2 = y'^2(0) \tag{3.4}$$

in a clockwise direction. See Fig. 1. The assumption $a\ell < V < b\ell$

implies that $\rho(x)$ is not identically equal to H. Therefore, there will come a time $0 < x_1 < \ell/2$ when the condition $\eta^{3/4}(y(x_1))^{-3/2} = H$ is satisfied. Then for sufficiently small values of $x > x_1$, we must have

$$\rho(x) = \eta^{3/4}(y(x))^{-3/2} .$$

For these values of x the differential equation (2.11) becomes

$$y'' + \lambda\eta^{3/4}y^{-1/2} = 0$$

and the point $(y(x),y'(x))$ is moving clockwise along the curve

$$(3.5) \qquad y'^2 + 4\lambda\eta^{3/4}y^{1/2} = y'^2(x_1) + 4\lambda\eta H^{-1/3} .$$

If h is sufficiently small, all points on this curve will satisfy $\eta^{3/4}y^{-3/2} > h$. We shall assume this to be the case. This amounts to the assumption that the optimal column nowhere achieves the maximum allowable thickness b. We leave to the reader the problem of determining ρ^* in the case where the maximum allowable thickness is achieved by the optimal column.

In the case we are considering the point $(y(x),y'(x))$ moves along the curve (3.5) on the interval $(x_1,\ell-x_1)$, and along the curve (3.4) on the interval $[\ell-x_1,\ell]$, as is indicated by the arrows in Fig. 1. Clearly, $y'(\ell/2) = 0$. Moreover, since eigenfunctions are unique only up to a scalar multiple we may assume that y has been scaled so that $y(\ell/2) = 1$. Equation (3.5) must then be given by

$$(3.6) \qquad y'^2 + 4\lambda\eta^{3/4}y^{1/2} = 4\lambda\eta^{3/4} .$$

Solving this equation for $y'(x_1)$ and substituting into (3.4) gives the equation

$$(3.7) \qquad y'^2 + \lambda H y^2 = 4\lambda\eta^{3/4} - 3\lambda\eta H^{-1/3}$$

for $(y(x),y'(x))$, $0 \le x \le x_1$. This equation can be solved for y in terms of x on this interval the solution shows that

$$(3.8) \qquad x_1 = \frac{1}{\sqrt{\lambda H}} \arcsin \sqrt{\frac{\eta H^{-4/3}}{4\eta^{3/4}H^{-1}-3\eta H^{-4/3}}} .$$

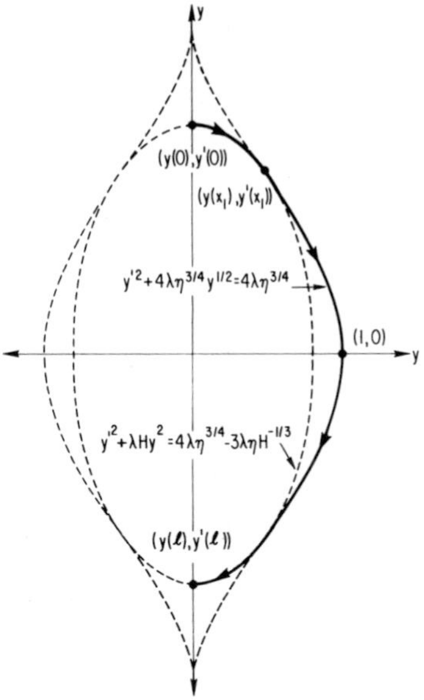

Figure 1. The path of the solution is indicated by arrows.

In order to obtain a second expression for x_1 we introduce a new function $\theta(x)$ defined by the requirement

$$y(x) = \sin^4\theta(x), \quad x_1 \leq x \leq \ell/2 .$$

Substituting this into (3.6) gives the differential equation

$$\sin^3\theta \frac{d\theta}{dx} = \frac{1}{2}\sqrt{\lambda\eta^{3/4}} , \quad x_1 \leq x \leq \ell/2$$

$$\theta(x_1) = \text{arc sin } (\eta^{1/8}H^{-1/6}) , \quad \theta(\ell/2) = \pi/2 ,$$

for θ. This implies that

$$\text{(3.9)} \qquad x_1 = \ell/2 - \frac{2}{\sqrt{\lambda\eta^{3/4}}} \int_{\text{arc sin } (\eta^{1/8}H^{-1/6})}^{\pi/2} \sin^2\theta d\theta .$$

Let $Z = \eta^{1/8}H^{-1/6}$. (3.8) and (3.9) then imply the equation

$$(3.10) \qquad \frac{1}{\sqrt{\lambda H}} \text{arc sin} \sqrt{\frac{Z^2}{4-3Z^2}} + \frac{2}{Z^3\sqrt{\lambda H}} \int_{\text{arc sin } Z}^{\pi/2} \sin^3\theta d\theta = \ell/2$$

for the unknowns Z and λ . A second equation for these unknowns is implied by the condition

$$\int_0^{\ell/2} \rho^{-1/3}(x)dx = V/2 .$$

We have

$$\int_0^{\ell/2} \rho^{-1/3}(x)dx = \int_0^{x_1} H^{-1/3}dx + \int_{x_1}^{\ell/2} \eta^{-1/4}\sin^2\theta dx$$

$$= \frac{H^{-1/3}}{\sqrt{\lambda H}} \text{arc sin} \sqrt{\frac{Z^2}{4-3Z^2}} + \frac{2H^{-1/3}}{Z^5\sqrt{\lambda H}} \int_{\text{arc sin } Z}^{\pi/2} \sin^5\theta d\theta = V/2 .$$

If we divide this equation by (3.10) we obtain the single equation

$$(3.11) \qquad \frac{\text{arc sin}\sqrt{\frac{Z^2}{2-Z^2}} + 2Z^{-5}\int_{\text{arc sin } Z}^{\pi/2} \sin^5\theta d\theta}{\text{arc sin}\sqrt{\frac{Z^2}{2-Z^2}} + 2Z^{-3}\int_{\text{arc sin } Z}^{\pi/2} \sin^3\theta d\theta} = V/a\ell$$

for Z . It is easily shown that this equation has a unique solution in the range $0 < Z < 1$.

Having determined Z , η can be obtained from the equation $Z = \eta^{1/8}H^{-1/6}$. If this value of η satisfies $\eta^{3/4} > h$, then the optimal column never achieves the maximum allowable thickness. In this case the appropriate value for λ can be obtained from equation (3.10). Given these values of η and λ , the cross sections $A(x)$ of the strongest thin-walled tubular column are obtained as follows. First solve the initial value problem

$$y'' + \lambda\rho(y)y = 0 , \quad 0 \leq x \leq \ell$$

$$y(0) = 0 , \quad y'(0) = \sqrt{4\lambda\eta^{3/4} - 3\lambda\eta H^{-1/3}} ,$$

where

$$\rho(y) = \begin{cases} H & \text{if } \eta^{3/4}y^{-3/2} \geq H \\ \eta^{3/2}y^{-3/2} & \text{if } \eta^{3/4}y^{-3/2} \leq H . \end{cases}$$

$A(x)$ is given by

$$A(x) = \begin{cases} a , & 0 \leq x \leq x_1 \\ \eta^{1/2}y^{1/2}(x) , & x_1 \leq x \leq \ell - x_1 \\ a , & \ell - x_1 \leq x \leq \ell \end{cases}$$

where x_1 is given by (3.8). The case where the optimal column achieves the maximum allowable thickness can be treated in an analogous manner.

4. Lyapunov Zones of Stability for Hill's Equation

We use the term "Hill's equation" to mean any homogeneous, linear, second-order differential equation with real, periodic coefficients. For the present discussion we restrict our attention to equations of the form

$$(4.1) \qquad y'' + (\lambda - \beta p(x))y = 0 , \quad -\infty < x < \infty$$

where λ and β are real parameters and p is a piecewise continuous function satisfying $p(x+\ell) = p(x)$ for all x. It is known that if λ and β are properly chosen, then all solutions of this equation will be bounded. On the other hand, an improper choice of λ and β will result in all nontrivial solutions being unbounded. For example, if $\beta = 0$ then all solutions are bounded for $\lambda > 0$ and all nontrivial solutions are unbounded for $\lambda > 0$. For $\lambda = 0$ the equation has a constant, hence periodic, solution. Our main purpose in this section is to determine conditions on λ and β that will guarantee the boundedness of all solutions. We begin with an example to motivate these considerations.

Consider a thin uniform rod of length L, undergoing small transverse vibrations due to periodic end forces $p(t) = p(t+\ell)$, $-\infty < t < \infty$. Let the line segment $0 \leq \xi \leq L$ represent the position of the rod in its undeflected state. In its deflected state, let $w(\xi,t)$ denote the deflection at ξ at time t. Under very mild assumptions (cf. [12]) w satisfies the differential equation

$$(4.2) \qquad EI \frac{\partial^4 w}{\partial \xi^4} + p(t) \frac{\partial^2 w}{\partial \xi^2} + m \frac{\partial^2 w}{\partial t^2} = 0 ,$$

where E is Young's modulus of the column material and I is the moment of inertia of cross sections. m is the mass per unit length of the rod. We assume that the rod has hinged ends so that in addition to (4.2) we have the boundary conditions

$$(4.3) \qquad w(0,t) = \frac{\partial^2 w}{\partial \xi^2}(0,t) = 0 , \quad w(L,t) = \frac{\partial^2 w}{\partial \xi^2}(L,t) = 0 .$$

Stability of the rod would require that all solutions of (4.2), (4.3) be bounded. In studying the boundedness of solutions of this boundary problem, it is customary to restrict attention to solutions of the form

$$(4.4) \qquad w(\xi,t) = \sum_{k=1}^{N} y_k(t) \sin \frac{k\pi\xi}{L} ,$$

that is, to solutions which have a finite Fourier sine series representation. By substituting this solution into (4.1) it is found that y_k satisfies

$$(4.5) \qquad m y_k''(t) + \left(EI\left(\frac{k\pi}{L}\right)^4 - \left(\frac{k\pi}{L}\right)^2 p(t)\right) y_k(t) = 0 , \quad k = 1,\dots,N .$$

This equation is of the form (4.1). The solutions (4.4) are bounded for arbitrary initial conditions if and only if all solutions of (4.5) are bounded. We shall determine stability conditions for λ and β in (4.1). If these conditions are satisfied for $\lambda = EI(\frac{k\pi}{L})^4$ and $\beta = (\frac{k\pi}{L})^2$, $k = 1,\dots,N$, then we can assert that solutions of (4.5) are bounded for all t.

Numerous applications requiring a stability analysis of equation (4.1) are discussed in [13]. The stability question itself has been studied by many authors. Good summaries of what has been done are given in [14] and [15]. The most fundamental results were given by Lyapunov [16] in 1899. For a modern treatment of these ideas see [15, Chapters 2 and 3]. We shall now describe the part of this theory that is required for our work.

For a fixed $\beta > 0$, let $\Lambda_1(\tau) < \Lambda_2(\tau) < \ldots$ denote the successive eigenvalues of the boundary-value problem

$$(4.6)\qquad \begin{aligned} &y'' + (\lambda - \beta p(x+\tau))y = 0 \\ &y(0) = y(\ell) = 0 \, . \end{aligned}$$

Let

$$\Lambda_n' = \min \Lambda_n(\tau) \, , \quad \Lambda_n'' = \max \Lambda_n(\tau) \, , \quad 0 \leq \tau \leq \ell \, ,$$

$n = 1,2,\ldots,$ and let

$$(4.7)\qquad \Lambda_0'' = \min \frac{\int_0^\ell [y'^2 + \beta p(x) y^2] dx}{\int_0^\ell y^2 dx}$$

where this latter minimization is taken over all absolutely continuous functions y satisfying $y(0) = y(\ell)$ and $y'(0) = y'(\ell)$. Then

$$\Lambda_0'' < \Lambda_1' \leq \Lambda_1'' < \Lambda_2' \leq \Lambda_2'' < \Lambda_3' \leq \Lambda_3'' < \ldots \, .$$

We shall sometimes use the notation $\Lambda_n'(\beta)$, $\Lambda_n''(\beta)$ to indicate the dependence of the numbers Λ_n', Λ_n'' on β.

Theorem 4.1. For a fixed β, all solutions of (4.1) are bounded if λ lies in one of the intervals $(\Lambda_n''(\beta), \Lambda_{n+1}'(\beta))$, $n = 0,1,\ldots$. All solutions are unbounded if λ lies in one of the intervals $(\Lambda_n'(\beta), \Lambda_n''(\beta))$, $n = 1,2,\ldots$. If λ is the endpoint of one of these intervals, then (4.1) has at least one nontrivial periodic, hence bounded, solution.

This theorem follows from Theorem 3.1.3 in [15]. The intervals $(\Lambda_n'', \Lambda_{n+1}')$ are called stability intervals, and the intervals $(\Lambda_n', \Lambda_n'')$ are called instability intervals.

Theorem 4.2. As $k \to \infty$, both Λ_{2k+1}' and Λ_{2k+1}'' satisfy

$$\sqrt{\Lambda} = (2k+1)\pi/\ell + (2k+1)^{-1}\pi^{-1}/2 \int_0^\ell p(x)dx + o(k^{-1})$$

and both Λ'_{2k+2} and Λ''_{2k+2} satisfy

$$\sqrt{\Lambda} = 2(k+1)\pi/\ell + (k+1)^{-1}\pi^{-1}/4 \int_0^\ell p(x)dx + o(k^{-1}) .$$

This theorem appears as Theorem 4.2.3 in [15]. It implies that the length of the n-th instability interval tends to zero as $n \to \infty$.

Consider problem (4.6). Theorem 4.2 shows that the asymtotic behavior of the eigenvalues depend only on $\int_0^\ell p(x)dx$. We remark, however, that the asymtotic convergence may be very slow. The asymtotic behavior of the eigenvalues of (4.6) suggests the consideration of the boundary-value problem

$$(4.8) \quad \begin{aligned} & y'' + (\lambda-\beta p(x))y = 0 , \quad 0 \leq x \leq \ell \\ & y(0) = y(\ell) = 0 , \end{aligned}$$

and the associated extremal eigenvalues

$$(4.9) \quad \lambda'_n = \min \lambda_n(\rho) , \quad n = 1,2,...,$$

$$(4.10) \quad \lambda''_n = \max \lambda_n(\rho) , \quad n = 1,2,...,$$

where the min and max are taken subject to

$$(4.11) \quad \int_0^\ell \rho(x)dx = M \equiv \int_0^\ell p(x)dx , \quad h \leq \rho(x) \leq H ,$$

where $h = \min p(x)$ and $H = \max p(x)$, $0 \leq x \leq \ell$. We also define $\lambda''_0 = \beta M$. By taking y constant in (4.7) we see that $\Lambda''_0 \leq \lambda''_0$. Since $\int_0^\ell p(x+\tau)dx = M$, $h \leq p(x+\tau) \leq H$, the following Theorem is clear.

Theorem 4.3. The interval $(\lambda''_n, \lambda'_{n+1})$ lies inside the stability interval $(\Lambda''_n, \Lambda'_{n+1})$, $n = 0,1,...$. Moreover, for sufficiently large values of n , the interval $(\lambda''_n, \lambda'_{n+1})$ is not empty.

We shall now give a brief indication of how the results of Section 2 can be used to compute the values λ'_n, λ''_n , $n = 1,2,...$. First, consider λ''_1 . Let ρ^* be a measurable function satisfying (4.11) and $\lambda_1(\rho^*) = \lambda''_1$. Let y^* be the corresponding eigenfunction of (4.8).

Then according to condition (2.24) of Theorem 2.3, there exist constants $\eta_0 \geq 0$, η such that

$$(4.12) \qquad \max_{h \leq \rho \leq H} \{\eta_0 y^{*2}(x)\rho - \eta\rho\} = \eta_0 y^{*2}(x)\rho^*(x) - \eta\rho^*(x) .$$

If $h\ell < M < H\ell$, then it is easy to show that $\eta_0 > 0$ and $\eta > 0$.

There are two possibilities for ρ^*. If the set of points x satisfying $y^{*2}(x) = \eta$ has measure zero, then (4.12) implies that ρ^* achieves only the values h and H. However, if $y^{*2}(x) = \eta$ on an interval of positive measure, then we must have $y^{*\prime\prime}(x) \equiv 0$, and consequently $\rho^*(x) \equiv \beta^{-1}\lambda_1(\rho^*)$, on this interval. Both possibilities are taken care of by the formula

$$(4.13) \qquad \rho^*(x) = \begin{cases} h \quad \text{if} \quad y^{*2}(x) < \sqrt{\eta} \\ \min\{\beta^{-1}\lambda_1(\rho^*), H\}, \quad y^{*2}(x) \geq \sqrt{\eta} . \end{cases}$$

If $\beta^{-1}\lambda_1(\rho^*) \leq H$, a simple phase plane analysis similar to that used in Section 2 shows the existence of a point $0 < x_1 < \ell/2$ such that

$$y^*(x) = \begin{cases} \dfrac{\sin\sqrt{\lambda-\beta h}\, x}{\sqrt{\lambda-\beta h}}, & 0 \leq x \leq x_1 \\ \sqrt{\eta}, & x_1 \leq x \leq \ell/2 . \end{cases}$$

The condition $\int_0^\ell \rho^*(x)dx = M$ implies that

$$x_1 = \frac{\lambda\ell-\beta M}{2(\lambda-\beta h)} .$$

The condition $y^{*\prime}(x) = 0$ holds for $x_1 \leq x \leq \ell/2$, and the condition $y^{*\prime}(x_1) = 0$ implies that $\lambda_1'' = \lambda_1(\rho^*)$ is the smallest solution of the equation

$$\sqrt{\lambda-\beta h}\, \frac{\lambda\ell-\beta M}{\lambda-\beta h} = \pi$$

in the range $\beta h < \lambda \leq \beta H$.

On the other hand, if it turns out that $\beta^{-1}\lambda_1(\rho^*) > H$, a direct substitution of (4.13) into (4.8) and solving the resulting equation, yields the equation

$$\sqrt{\lambda-\beta h}\cot\sqrt{\lambda-\beta h}\,x_1 = \sqrt{\lambda-\beta H}\tan\sqrt{\lambda-\beta H}\,(\ell/2-x_1)$$

for $\lambda_1'' = \lambda_1(\rho^*)$, where

$$x_1 = \frac{\ell H-M}{2(H-h)} .$$

To find λ_n'' for $n > 1$ divide the interval $[0,\ell]$ into n equal parts and consider the boundary-value problem

$$(4.14)\qquad \begin{aligned} &y'' + (\lambda-\beta\rho(x))y = 0 , \quad 0 \leq x \leq \ell/n \\ &y(0) = y(\ell/n) = 0 , \end{aligned}$$

and the constraints

$$(4.15)\qquad \begin{aligned} &\int_0^{\ell/n} \rho(x)dx = M/n , \\ &h \leq \rho(x) \leq H . \end{aligned}$$

Maximize the first eigenvalue of (4.14) subject to (4.15). Let ρ^* denote the solution and let y^* denote the corresponding eigenfunction. Extend ρ^* periodically to the entire interval $[0,\ell]$ by defining $\rho^*(x+\ell/n) = \rho^*(x)$. Extend y^* semi-periodically by defining $y^*(x+\ell/2) = -y^*(x)$. Clearly, the extended functions satisfy condition (2.24) of Theorem 2.3. Moreover, it can be shown that these functions are in fact optimal.

In summary we have the following:

Theorem 4.4. For $n = 1,2,\ldots$, λ_n'' , defined by (4.10), is the smallest solution of the equation

$$\frac{\lambda\ell-\beta M}{\sqrt{\lambda-\beta h}} = n\pi$$

in the range $\beta h < \lambda \leq \beta H$ provided such exists, and is otherwise the smallest solution of the equation

$$\sqrt{\lambda-\beta h}\cot\sqrt{\lambda-\beta h}\,x_1/n = \sqrt{\lambda-\beta H}\tan\sqrt{\lambda-\beta H}\,(\ell/2-x_1)/n$$

in the range $\lambda > \beta H$, where $x_1 = \frac{\ell H-M}{2(H-h)}$.

A similar result holds regarding the numbers λ_n' defined in (4.9). We shall simply state the result in the next theorem and leave the proof to the reader.

Theorem 4.5. For $n = 1,2,\ldots$, λ_n' is the smallest solution of the equation

$$\sqrt{\beta H-\lambda}\ \coth \sqrt{\beta H-\lambda}\ x_1/n = \sqrt{\lambda-\beta h}\ \tan \sqrt{\lambda-\beta h}\ (\ell/2-x_1)/n$$

in the range $\beta h < \lambda \leq \beta H$, provided such exists, and is otherwise the smallest solution of the equation

$$\sqrt{\lambda-\beta H}\ \cot \sqrt{\lambda-\beta H}\ x_1/n = \sqrt{\lambda-\beta h}\ \tan \sqrt{\lambda-\beta h}\ (\ell/2-x_1)/n$$

in the range $\lambda > \beta H$. Here $x_1 = \frac{M-\ell h}{2(H-h)}$.

Example. All solutions of the equation

$$y'' + (\lambda-\cos 2x)y = 0 \tag{4.16}$$

are bounded if λ lies in one of the intervals

(0, .31541), (1.57577, 3.35029), (4.62272, 8.35746), (9.63054, 15.36003), (16.63322, 24.36123), (25.63445, 35.36189), (36.63511, 48.36228), (49.63551, 63.36254), (64.63578, 80.36271), (81.63595, 99.36285), (100.63607, 120.36294), ...

These are the intervals $(\lambda_n'', \lambda_{n+1}')$, $n = 0,\ldots,10$, obtained by applying Theorems 4.3, 4.4, 4.5, with $\beta = 1$, to equation 4.16.

The first application of optimal control theory to stability problems for Hill's equation was made by Brockett in [17].

5. A Variational Problem Arising in the Design of Cooling Fins

When it is desired to increase the heat removal from a structure to a surrounding medium, it is common practice to utilize extended surfaces attached to the primary surface. Examples may be found in the cooling fins of air-cooled engines, the fin extensions to the tubes of radiators, the pins or studs attached to boiler tubes, etc. In this section we shall study annular fins attached circumferentially to a cylindrical surface. See Fig. 2. The question we ask is this: Given a fin of fixed weight and length, and thickness $\geq h$ and $\leq H$, how should it be tapered in order to maximize the rate of heat dissipation to the surrounding medium. The answer was conjectured by Schmidt [18] in 1926 for fins with no minimum or maximum thickness constraint imposed on them. He proposed that the optimum fin should taper, narrowing in the direction of heat flow, in such a way that the gradient of the temperature in the fin is constant. This conjecture was proved rigorously by Duffin [19] in 1959. Since that time a number of papers have appeared treating various aspects of the optimal design problem. We list [20], [21], [22], [23], to name a few. In [23] the present author obtained the optimum taper of a rectangular fin subject to thickness constraints. We shall now show how to obtain analogous results for annular fins.

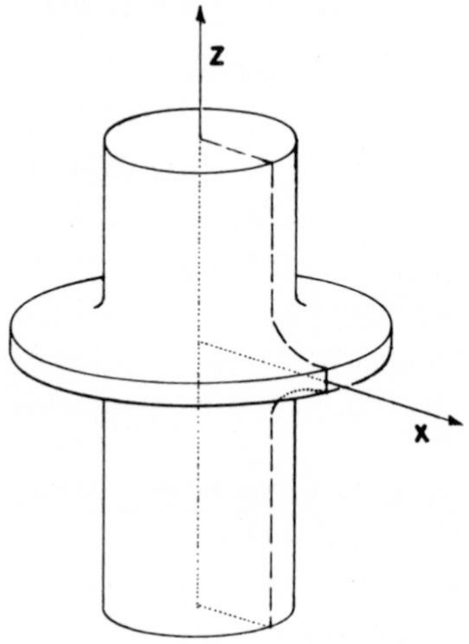

Figure 2. Annular Fin on a Cylinder

In x, y, z space we take the fin to be parallel to the x,y plane. We assume the fin is sufficiently thin that there is no appreciable change in its temperature in the z direction. If the temperature of the surrounding medium is taken to be zero, and if we assume that Newton's thermal conductivity is unity, then the flow of heat in the fin is governed by the steady state heat equation

$$\frac{\partial}{\partial x}\left(p\,\frac{\partial u}{\partial x}\right) + \frac{\partial}{\partial y}\left(p\,\frac{\partial u}{\partial y}\right) - qu = 0, \quad (x,y) \in S \,, \tag{5.1}$$

where S is the annular region of the fin in the x,y plane. $u = u(x,y)$ is the temperature in the fin, $p = p(x,y)$ is the thickness of the fin, and $q > 0$ is the cooling coefficient, here assumed constant. We shall assume that the outer edge of the fin is insulated so that the appropriate boundary conditions are

$$\begin{aligned} &u = T \text{ (= steady state temperature) on } \Gamma_1 \\ &\frac{\partial u}{\partial \nu} = 0 \quad \text{on } \Gamma_2 \end{aligned} \tag{5.2}$$

where Γ_1 and Γ_2 are, respectively, the inner and outer boundaries of S . ν is a unit outward normal to the boundary of S .

Newton's law of cooling implies that the heat dissipated per unit time by the fin is given by

$$\iint_S q\, u\, dxdy \,. \tag{5.3}$$

The weight of the fin, which we assume to be fixed, is proportional to

$$\iint_S p(x,y)dxdy \equiv M \,. \tag{5.4}$$

The thickness of the fin satisfies the constraints

$$h \leq p(x,y) \leq H \tag{5.5}$$

for some positive numbers h and H satisfying

$$h \iint_S dx\, dy < M < H \iint_S dx\, dy \,.$$

Our problem is to determine p and u satisfying conditions (5.1), (5.2), (5.4), and (5.5), in such a way that the integral (5.3) is maximized. Duffin and McLain [24] have given a max-min formulation of this problem. In order to obtain this formulation multiply equation (5.1) by u and integrate, using Green's theorem, to obtain

$$\iint_S [p\,(\frac{\partial u}{\partial x})^2 + p\,(\frac{\partial u}{\partial y})^2 + qu^2]\,dxdy$$

(5.6)
$$= \iint_S [qu - \frac{\partial}{\partial x}(p\,\frac{\partial u}{\partial x}) - \frac{\partial}{\partial y}(p\,\frac{\partial u}{\partial x})]udxdy + \int_{\Gamma_1} up\,\frac{\partial u}{\partial \nu}\,d\sigma + \int_{\Gamma_2} up\,\frac{\partial u}{\partial \nu}\,d\sigma$$

$$= T\int_{\Gamma_1} p\,\frac{\partial u}{\partial \nu}\,d\sigma\ .$$

On the other hand, by simply integrating equation (5.1) over the region S, again making use of Green's theorem, we obtain

(5.7)
$$\iint_S qudxdy = \iint_S [\frac{\partial}{\partial x}(p\,\frac{\partial u}{\partial x}) + \frac{\partial}{\partial y}(p\,\frac{\partial u}{\partial y})]\,dxdy = \int_{\Gamma_1} p\,\frac{\partial u}{\partial \nu}\,d\sigma\ .$$

Combining this with (5.6) we obtain

(5.8)
$$\iint_S qudxdy = \frac{1}{T}\iint_S [p\,\frac{\partial u}{\partial x})^2 + p\,(\frac{\partial u}{\partial y})^2 + qu^2]\,dxdy\ .$$

The differential equation (5.1), together with the boundary condition $u = T$ on Γ_1 is just the Euler equation for minimizing the integral on the right in (5.8). We can therefore write

(5.9)
$$\iint_S qudxdy = \frac{1}{T}\min_v \iint_S [p\,(\frac{\partial v}{\partial x})^2 + p\,(\frac{\partial v}{\partial y})^2 + qv^2]\,dxdy$$

where the minimization is taken subject to $v = T$ on Γ_1. The problem of tapering the fin to maximize the rate of heat dissipation is therefore equivalent to the max-min problem

$$(5.10) \qquad \max_p \min_u \iint_S \left[p\left(\frac{\partial u}{\partial x}\right)^2 + p\left(\frac{\partial u}{\partial y}\right)^2 + qu^2\right] dxdy$$

where the minimization is taken over functions u satisfying $u = T$ on Γ_1 and the maximization is taken over functions p satisfying (5.4) and (5.5).

The problem we have formulated is valid for fins on cylinders of arbitrary cross-sectional type. However, we shall now restrict our attention to circular cylinders of radius R and fins of length ℓ. In this case it is clear that the functions p and u which solve (5.10) must depend only on the distance to the center of the cylinder.

Let the center line of the cylinder be along the z axis as in Fig. 2. Introduce the variables $r = \sqrt{x^2+y^2}$, $\phi(r) = u(x,y)$, $\rho(r) = p(x,y)$. In terms of these variables the problem (5.10) becomes

$$(5.11) \qquad \max_\rho \min_\phi \int_R^{R+\ell} [r\rho(r)\phi'^2(r) + qr\phi^2(r)]dr$$

where the minimization is taken over absolutely continuous functions ϕ satisfying

$$(5.12) \qquad \phi(R) = T$$

and the maximization is taken over piecewise continuously differentiable functions $\rho(r)$ satisfying

$$(5.13) \qquad \int_R^{R+\ell} r\rho(r)dr = \frac{M}{2\pi},$$

$$h \le \rho(r) \le H .$$

The boundary-value problem (5.1), (5.2) transforms into

$$\frac{d}{dr}\left(r\rho(r)\frac{d\phi}{dr}\right) - qr\phi(r) = 0 \tag{5.14}$$
$$\phi(R) = T,\ \phi'(R+\ell) = 0\ .$$

The technique used to prove condition (3.14) in [23] can be used to prove the following theorem.

Theorem 5.1. Let ρ^* and ϕ^* be functions satisfying (5.13) and (5.14). The pair (ρ^*,ϕ^*) is a solution of (5.11) if and only if there exists a constant $\eta > 0$ such that

$$\max_{h \le \rho \le H} \rho[\phi^{*\prime 2}(r)-\eta] = \rho^*(r)[\phi^{*\prime 2}(r)-\eta] \tag{5.15}$$

for each r in $[R, R+\ell]$.

Let (ρ^*,ϕ^*) be a solution of (5.11). Since $\phi^{*\prime}(R+\ell) = 0$, condition (5.15) implies that $\rho^*(r) = h$ for values of r sufficiently close to $R+\ell$. For these values of r , equation (5.14) is of Bessel type. It therefore seems unreasonable to attempt an analytic solution of (5.11). Instead we shall give an iterative procedure which can be used to obtain numerical solutions. First we remark that in case there are no constraints of the form $h \le \rho(r) \le H$ on ρ, the optimal ϕ satisfies $\phi'^2(r) = \eta$, $R \le r \le R+\ell$. Substituting this into (5.14) gives a simple differential equation from which ρ can be determined. This analysis is carried out in [19].

To facilitate the discussion of a numerical solution of (5.11) we introduce some notation. All functions involved will be considered as members of $L_2[R, R+\ell]$, the space of square integrable functions on $[R, R+\ell]$. We shall use the symbol $\phi\cdot\rho$ to denote the inner product $\int_0^\ell \phi(r)\rho(r)dr$ of two functions in $L_2[R, R+\ell]$. The norm in this space will be denoted as usual, by $\|\ \|$. For convenience, we define a functional g on the class of nonnegative ρ's by

$$g(\rho) = \min_{\phi} \int_R^{R+\ell} [r\rho(r)\phi'^2(r) + qr\phi^2(r)]dr \tag{5.16}$$

where the minimization is taken over functions satisfying $\phi(R) = T$. Problem (5.11) is then to maximize g subject to (5.13).

Let (ρ_1,ϕ_1), (ρ_2,ϕ_2) be pairs of functions satisfying (5.14). Then by arguing as in the proof of Theorem 3.2 in [23], it can be shown that

$$(5.17)\qquad \int_0^{\ell} r\phi'^2(\rho_2-\rho_1)dr - \frac{K}{2}\int_0^{\ell}(\rho_2-\rho_1)^2 dr$$

$$\leq g(\rho_2) - g(\rho_1) \leq \int_0^{\ell} r\phi'^2(\rho_2-\rho_1)dr ,$$

where $K = \{\sqrt{R+\ell}\,\frac{qT}{Rh}(2R+\ell)\ell\}^2/h$. This means that g is differentiable, and its gradient is given by

$$\nabla g(\rho) = r\phi'^2$$

where ϕ is the function which solves the minimization problem (5.16).

Let ρ^* be the function which maximizes g subject to (5.13). A sequence of functions converging to ρ^* can be generated as follows.

i) Let ρ_1 satisfying (5.13) be chosen arbitrarily.

ii) If $\rho_1,\ldots,\rho_k$ have been chosen, take ρ_{k+1} to be the solution of

$$\text{maximize} \int_R^{R+\ell} [\nabla g(\rho_k)\cdot(\rho-\rho_k) - \frac{K}{2}(\rho-\rho_k)^2]dr$$

subject to

$$(5.18)\qquad \int_R^{R+\ell} r\rho(r)dr = \frac{M}{2\pi} ,$$

$$h \leq \rho(r) \leq H .$$

ρ_{k+1} is the solution of a simple moment problem which can be easily solved numerically. It is clear from (5.17) that $g(\rho_{k+1}) \geq g(\rho_k)$. We shall now show that the sequence $g(\rho_1)$, $g(\rho_2)$, ..., actually converges to $g(\rho^*)$.

By completing the square in the integrand in ii) above, we see that ρ_{k+1} is the point in the convex constraint set defined by (5.18), nearest the point $\rho_k + \frac{1}{K}\nabla g(\rho_k)$. It follows that

$$(5.20) \qquad (\rho_k + \frac{1}{K} \nabla g(\rho_k) - \rho_{k+1}) \cdot (\rho - \rho_{k+1}) \leq 0$$

for any ρ satisfying the constraints in (5.18), and in particular for $\rho = \rho_{k+1}$.

For $k \geq 1$, (5.17) implies that

$$\begin{aligned} g(\rho_{k+1}) - g(\rho_k) &\geq \nabla g(\rho_k) \cdot (\rho_{k+1} - \rho_k) - \frac{K}{2} \|\rho_{k+1} - \rho_k\|^2 \\ &= K(\rho_k + \frac{1}{K} \nabla g(\rho_k) - \rho_{k+1}) \cdot (\rho_{k+1} - \rho_k) \\ &\quad + K\|\rho_{k+1} - \rho_k\|^2 - \frac{K}{2} \|\rho_{k+1} - \rho_k\|^2 \\ &\geq \frac{K}{2} \|\rho_{k+1} - \rho_k\|^2 \qquad \text{(by (5.20))}. \end{aligned}$$

Since $\sum_{k=1}^{\infty} [g(\rho_{k+1}) - g(\rho_k)] \leq g(\rho^*) - g(\rho_1)$, it follows that

$$(5.21) \qquad \lim_{k \to \infty} \|\rho_{k+1} - \rho_k\| = 0 .$$

(5.17) implies that

$$\begin{aligned} 0 \leq g(\rho^*) - g(\rho_k) &\leq \nabla g(\rho_k) \cdot (\rho^* - \rho_k) \\ &= \nabla g(\rho_k) \cdot (\rho_k - \rho_{k+1}) + K(\rho_k + \frac{1}{K} \nabla g(\rho_k) - \rho_{k+1}) \cdot (\rho^* - \rho_{k+1}) \\ &\qquad - K(\rho_k - \rho_{k+1}) \cdot (\rho^* - \rho_{k+1}) \\ &\leq (\nabla g(\rho_k) - K(\rho^* - \rho_{k+1})) \cdot (\rho_k - \rho_{k+1}) \to 0 \end{aligned}$$

as $k \to \infty$ by (5.21). This completes the proof that $\lim_{k \to \infty} g(\rho_k) = g(\rho^*)$. Thus for sufficiently large k , the efficiency of ρ_k is close to that of ρ^* .

For a different treatment of problem (5.9) see [25].

ACKNOWLEDGEMENT

I acknowledge with pleasure the assistance of my colleague Matthew Halfant. He drew the figures and assisted with some computer programming required to test the validity of certain results.

REFERENCES

1. J. L. de Lagrange, "Sur la figure des colonnes, Miscellana Taurinensia," Tomus 5, 1770-1773, p. 123.
2. T. Clausen, "Über die Form architektonischer Säulen," Bulletin physicomathématiques et Astronomiques, Tome 1, 1849-1853, pp. 279-294.
3. I. Tadjbakhsh and J. B. Keller, "Strongest columns and isoperimetric inequalities for eigenvalues," ASME Journal of Applied Mechanics, Vol. 29, Series E, 1962, pp. 159-164.
4. F. I. Niordson, "On the optimal design of a vibrating beam," Quarterly of Applied Mathematics, Vol. 23, No. 1, 1965, pp. 47-53.
5. B. Budiansky, J. C. Frauenthal, J. W. Hitchinson, "On optimal arches," Transactions of the ASME, Series E, December 1969, pp. 880-882.
6. W. Prager and J. E. Taylor, "Problems of optimal structural design," J. Applied Mechanics, Vol. 35, 1968, pp. 102-106.
7. J. E. Taylor, "The strongest column: An energy approach," J. Applied Mechanics, Vol. 34, No. 2, Transactions of the ASME, Vol. 89, Series E, 1967, p. 486.
8. Journal of Optimization Theory and Applications, Vol. 15, No. 1, January 1975.
9. M. Feigen, "Minimum weight of tapered round thin-walled columns," J. Applied Mechanics, Vol. 19, Transactions ASME, Vol. 74, 1952, pp. 375-380.
10. E. R. Barnes, "The shape of the strongest column and some related extremal eigenvalue problems," to appear in Quarterly of Applied Mathematics.
11. M. R. Hestenes, Calculus of Variations and Optimal Control Theory, John Wiley and Sons, Inc., New York, 1966.
12. S. Lubkin and J. J. Stoker, "Stability of columns and strings under periodically varying forces," Quarterly of Applied Mathematics, Vol. 1, No. 3, 1943, pp. 215-236.
13. V. V. Bolotin, "The Dynamic Stability of Elastic Systems," Holden Day, Inc., San Francisco, 1964.
14. W. Magnus and S. Winkler, "Hill's Equation," Interscience Publishers, New York, 1966.
15. M. S. P. Eastham, "The Spectral Theory of Periodic Differential Equations," Hafner Press, New York, 1973.
16. A. M. Lyapunov, "Sur une équation différentielle linéaire du second ordre, C. R. Acad. Sci. Paris 128, 1899, pp. 910-913.
17. R. W. Brockett, "Variational methods for stability of periodic equations," in Differential Equations and Dynamical Systems, Hale and LaSalle, editors, Academic Press, New York, 1967.
18. E. Schmidt, "Die wärmeübertragung durch rippen," Zeit. d. Ver. Deutch Ing. 70, 1926, pp. 885-890.
19. R. J. Duffin, "A variational problem relating to cooling fins," J. Mathematics and Mechanics, Vol. 8, 1959, pp. 47-56.
20. J. E. Wilkens, Jr., "Minimum mass thin fins and constant temperature gradients, SIAM J., Vol. 10, 1962, pp. 62-73.
21. C. Y. Lin, "A variational problem with applications to cooling fins," SIAM J., Vol. 10, 1962, pp. 19-29.

22. C. Y. Lin, "A variational problem relating to cooling fins with heat generation," Quarterly of Applied Mathematics, Vol. 19, 1962, pp. 245-251.
23. E. R. Barnes, "A variational problem arising in the design of cooling fins," Quarterly of Applied Mathematics, Vol. 34, No. 1, April, 1976, pp. 1-17.
24. R. J. Duffin and D. K. McLain, "Optimum shape of a cooling fin on a convex cylinder," J. Mathematics and Mechanics, Vol. 17, 1968, pp. 769-784.
25. J. Cea and K. Malanowski, "An example of a max-min problem in partial differential equations," SIAM J. Control, Vol. 8, No. 3, 1970, pp. 305-316.

"VARIATIONAL METHODS FOR THE NUMERICAL SOLUTIONS OF FREE BOUNDARY PROBLEMS AND OPTIMUM DESIGN PROBLEMS"

O. Pironneau

ABSTRACT

This paper is concerned with the resolution of partial differential equations with an additional boundary condition but an unknown boundary. Several methods are mentioned (variational inequalities, transformations of variables, relaxation methods, evolution methods) but the emphasis is put on the methods of optimum design and their numerical implementations.

PLAN

1. Introduction
2. Variational inequalities and transformations of variables
3. Relaxation methods and evolutions methods
4. The methods of optimum design
5. Conclusion

1. INTRODUCTION

Free boundary problems are boundary-value problems in which part of the boundary, the free boundary, is unknown and must be determined as a part of the problem. Such problems arise in continuum mechanics when two fluids are involved (ex. jets, waves...) or when a system with certain desired properties is to be designed (inverse problems in the design of airfoils, for example). It is beyond the scope of this paper to survey the work that has been done in this area. For one thing the task would be enormous since free boundary problems arise in many different fields. The oldest work that I have seen is the one mentioned in Lamb (1879) on the determination of the profile of minimum drag in

Stokes flow. Cryer (1970) mentions a paper by Trefftz (1916). Let me just say that a minimum references list on the subject should include the work of Hadamard (1910), Southwell (1946), Garabedian (1956), Lions (1972).

Among the various methods, that I know, for solving free boundary problems one should distinguish between the general methods and those which are designed specifically for certain problems. Among the general methods one must also distinguish between those which have theoretical proofs of convergences and those which do not. Thus we have divided the paper in three paragraphs accordingly, but since they are rather new, we shall speak mostly of the general methods of optimum design, which are known to converge.

2. VARIATIONAL INEQUALITIES AND TRANSFORMATIONS OF VARIABLES

2.1. Variational inequalities

Among the methods which are not general perhaps the best one is the method of variational inequalities. Let us consider the following problem: given $\Omega \subset R^n$; $a_{ij}(x)$, $a_0(x) \in L^\infty(\Omega)$, $a_{ij} = a_{ji}$, i, j=1, ..., n, $f \in L^2(\Omega)$, $g \in L^2(\Gamma)$, find ϕ such that:

$$\min_{\phi \in H^1(\Omega)} \{\int_\Omega \frac{1}{2} (a_{ij} \frac{\partial\phi}{\partial x_i} \frac{\partial\phi}{\partial x_j} + a_0\phi^2)dx - \int_\Omega f\phi dx - \int_\Omega g\phi d\Gamma \mid \phi \geq 0 \text{ a.e. in } \Omega\} \tag{2.1}$$

where

$$H^1(\Omega) = \{\phi \in L^2(\Omega) \mid \frac{\partial\phi}{\partial x_i} \in L^2(\Omega) \quad i=1,\ldots,n\} .$$

If there exists $\alpha > 0$ such that $a_{ij}\xi_i\xi_j \geq \alpha|\xi|^2$ $\forall\xi \in R^n$, $a_0 \geq \alpha$, then (2.1) has an unique solution and it is such that there exists Ω^+, Ω^0 with $\Omega^+ \cap \Omega^0 = \emptyset$, $\overline{\Omega^+ \cup \Omega^0} = \bar{\Omega}$.

(2.2) $$-\frac{\partial}{\partial x_i}\left(a_{ij}\frac{\partial \phi}{\partial x_j}\right) + a_0\phi = f \quad \text{in} \quad \Omega^+$$

(2.3) $$\frac{\partial \phi}{\partial \nu}\Big|_{\Gamma^+} = g \quad \text{or} \quad \phi\Big|_{\Gamma^+} = 0 \quad (\Gamma^+ = \partial\Omega^+, \text{ the boundary of } \Omega^+, \ \nu = a_{ij}\text{-co-normal})$$

(2.4) $$\phi|_S = 0 \quad \text{and} \quad \frac{\partial \phi}{\partial \nu}\Big|_S = 0 \qquad (S = \partial\Omega^+ \cap \partial\Omega^\circ)$$

(2.5) $$\phi = 0 \quad \text{in} \quad \Omega^\circ .$$

Therefore the free boundary problem: find (ϕ,Ω^+) such that (2.2)-(2.4) can be replaced by (2.1).

Problem (2.1) can be discretized by the method of finite elements and solved by numerical techniques related to those of variational equalities (see Glowinski-Lions-Tremolières(1976)).

Several free boundary problems have been solved successfully by this method: the determination of the plastic region for a Bingham fluid (Cea-Glowinski (1972)), the determination of the region of infiltration in a dam (Baiocchi-Comincolo-Guerri-Volpi (1973)), the determination of a cavitation bubble attached to an airfoil (Bourgat-Duvaut (1975)) and the determination of the surface of fusion of a two-phases problem (Bourgat-Duvaut (1976)).

This method gives good results for the following reasons:

1. The discretization need not follow the unknown boundary
2. The accuracy on ϕ does not depend on the accuracy of Ω^+ (see figure 1)
3. It is not necessary to know a good estimate of Ω^+ .

Unfortunately not all free boundary problems can be formulated as variational inequalities, or quasi-variational inequalities (Bensoussan-Lions (1973)). For a survey of the problems on which this method is feasible see Baiocchi (1973).

2.2. Transformation of variable

The essence of the method is to find a new statement of the problem by a change of variable and/or coordinates. The new problem will be either a non-linear partial differential equation with non-linear boundary conditions but fixed domain, or even a free boundary problem of the kind studied in §2.1 (Baiocchi (1973)). Let us just mention briefly a famous example: the inverse problem in incompressible flow, (see for example Stanitz (1952)) i.e. find the shape of the wall of a nozzle that gives a desired velocity. If the nozzle is symmetric and of infinite length, let $\phi(x,y)$ be the potential, $\psi(x,y)$ the stream function, given $\phi_0(s)$, ϕ_∞,k the problem is to find $S = (S_1,S_2)$ such that (see figure 2)

(2.6) $$\Delta\phi = 0 \quad \phi|_S = \phi_0(s) \ , \ \frac{\partial\phi}{\partial n}|_S = 0 \quad \phi|_\infty = \phi_\infty$$

or

(2.7) $$\Delta\psi = 0 \quad \psi|_{S_1} = 0 \ , \ \psi|_{S_2} = k \ , \ \frac{\partial\psi}{\partial n}|_S = \frac{\partial\phi_0}{\partial s} \ , \ \psi|_\infty = ky$$

Let $\theta(x,y)$ be the angle of the stream lines with the Ox axis, then it is easy to show that in the (ϕ,ψ) plane the nozzle becomes a stripe $(-\infty,+\infty) \times [0,k]$ in which

(2.8) $$\Delta\theta = 0$$

The continuity equation implies

(2.9) $$\frac{\partial\theta}{\partial\psi} = -\frac{1}{v}\frac{\partial v}{\partial\phi} \quad \text{on the boundaries} \quad \psi = 0, \ \psi = k \ ;$$

where $v(\phi)$ is obtained from the equations

(2.10) $$V(\phi) = \frac{\partial\phi_0}{\partial s}(s(\phi))$$

(2.11) $$\phi(s) = \phi_0(s)$$

(Ex: $\phi_0(s) = s^2 \Rightarrow V(\phi) = 2\sqrt{\phi}$)

Therefore the free boundary problem (2.6) has been replaced by a Neuman problem (2.7)-(2.8) plus a non-linear equation (2.9)-(2.10). Naturally this kind of technique is not general but when it works it is quite remarkable; for instance some inverse problems in transonic flow are easier than the computation of the solution of transonic equation with fixed boundaries.

3. RELAXATION METHODS AND EVOLUTION METHODS

3.1. Relaxation methods

Relaxation methods (or fixed point methods) were studied by Southwell (1956) and are widely used by engineers. The principle is the following: suppose we must find (S,ϕ) such that

(3.1) $$\Delta\theta = f \quad \text{in} \quad \Omega_S$$

(3.2) $$\phi|_\Gamma = h|_\Gamma \ , \ \Gamma = \Sigma \cup S = \partial\Omega \ , \Sigma \cap S = \emptyset$$

(3.3) $$\frac{\partial\phi}{\partial n}\Big|_S = g|_S$$

where f, h, g are given functions of $L^2(\Omega)$ and Σ is given. Then one can choose an initial guess S_0 for S and solve (3.1) with the boundary conditions

(3.4) $$\frac{\partial\phi}{\partial n}\Big|_{S_0} = g|_{S_0} \ , \ \phi|_\Sigma = h|_\Sigma$$

and compute S_1 such that $\phi|_{S_1} = h|_{S_1}$

This way it is possible to generate a sequence $\{S_i\}_{i\geq 0}$ which will (hopefully) converge to the desired solution. Refinements of this method were given by Garabedian (1956), Cryer (1970), ...

Relaxation methods are quite general and very easy to use; however, no proofs of convergence are available and in fact it is usually necessary to start with a good initial guess, otherwise the method diverges.

3.2. Evolution methods

It is a standard procedure in physics to replace a stationary problem by a time dependent problem which has the stationary problem for equilibrium state.

For example, the interface between two viscuous fluids in equilibrium in a rotating cylinder, under gravity, can be determined by starting with the fluids at rest, and slowly accelerate the cylinder to the desired speed, while following the interface. The position of the interface at time $t + dt$ is determined from the position at time t and the knowledge of the speed of the fluid particles.

In this example the time dependent problem is given by the physics of the problem, whenever the free boundary problem is a straightforward modelization of a physical problem, the time dependent problem is easy to construct.

Note that it is not fair to put these methods in this paragraph since proofs of convergence are indeed available for them, if they are properly applied. They are easy to use but they are rather slow and not always numerically stable, because they correspond to an explicit discretization scheme for a parabolic problem. A seepage problem in a dam has been solved numerically by this method (Zienkiewicz (1971)).

Cea (1974) showed that it is generally possible to formulate free boundary problems in this manner;however,the computation of the speed of the particles is not at all straightforward so that his method really belongs to the next paragraph.

4. THE METHODS OF OPTIMUM DESIGN

Consider problem (3.1)-(3.3) and let ϕ^S be the solution of

$$\Delta\phi^S = f \quad \text{in} \quad \Omega s, \quad \phi^S|_\Gamma = h|_\Gamma \tag{4.1}$$

and ψ^S the solution of

$$\Delta\psi^S = f \quad \psi^S|_\Sigma = h|_\Sigma \, , \; \frac{\partial\psi}{\partial n}\,|_S = g|_S \tag{4.2}$$

Then any solution S of (3.1)-(3.3) is also a solution of

(4.3) $$\min_{S \in \mathcal{S}} \{\int_{\Omega_S} |\nabla(\phi^S - \psi^S)|^2 dx\}$$

where $\mathcal{S}$ is the set of admissible boundaries S such that $\partial\Omega_S = \Sigma \cup S$, and (4.1)-(4.2) have solutions in Ω_S.

Problem (4.3) is an optimal control problem of a distributed parameter system where the (open loop) control is a part of the boundary; such a problem is a problem of optimum design.

Existence theorems for the solutions of (4.3) were given for special cases by Garabedian (1964) and Lions (1972).

It was shown by Murat-Simon (1974), Chesnay (1975), Pironneau (1976) that problems of the type of (4.3) with convex functional and linear operator* have solutions if ε contains uniformly Lipschitz boundaries (boundaries of bounded variations in R^2 is sufficient). The Lipschitz condition can even be removed in certain cases (Pironneau (1976)).

Problems (4.3) can be solved by the techniques of the calculus of variations and mathematical programming. For notational convenience let us illustrate the methods on two model problems

(4.4) $$\min_{S \in \mathcal{S}} \{\int_{\Omega} |\phi - \phi_d|^2 dx | \ -\Delta\phi + \phi = f \ , \ \frac{\partial\phi}{\partial n}\Big|_S = 0 \ , \ S = \partial\Omega\}$$

(4.5) $$\min_{S \in \mathcal{S}} \{\int_{\Omega} |\phi - \phi_d|^2 dx | \quad \Delta\phi = f, \ \ \phi|_S = 0 \ , \quad S = \partial\Omega\}$$

4.1. Reduction to a fixed domain

Following Begis-Glowinski (1973), (1976), Murat-Simon (1974), Morice (1974), let T_S^{-1} be a (smooth) mapping of R^2 into R^2 which transforms all admissible domains Ω_S, $S \in \mathcal{S}$ into a fixed domain Ω_0. Then problem (4.5), for instance, becomes

---

* These assumptions can be relaxed.

$$(4.6)\quad \min_{T\in\tau} \{\int_{\Omega_0} |\phi_T\ \phi_d|^2 |\det T'|d\Omega_0| \int_{\Omega_0} < {}^tT'^{-1}\nabla\phi_T\ ,\ {}^tT'^{-1}\nabla w > |\det T'|d\Omega_0$$

$$= \int_{\Omega_0} f_T\ w|\det T'|d\Omega_0\} \quad , \forall w \in H_0^1(\Omega_0)\ ,\ \phi_T \in H_0^1(\Omega_0)$$

where $\phi_T = \phi\circ T$, $H_0^1(\Omega_0) = \{\phi\in H^1(\Omega_0)|\phi|_{\Gamma_0} = 0\}$, $T' = \left(\frac{\partial T_i}{\partial x_j}\right)$, ${}^tT'$ = transpose of T', and τ is the set of all admissible mappings.

Problem (4.6) is a problem of optimal control where the control appears in the coefficient of the P.D.E. This type of problem was studied by Chavent (1971) (among others); they can be solved by the method of steepest descent (or conjugate gradient). However, the equations are rather complex as one might guess. This method has also the drawback of requiring the explicit knowledge of τ which for complicated domains is not a trivial matter. It has however the advantage of being easy to discretize.

Begis-Glowinski (loc. cit.) have tried this method on a simple geometry where the upper boundary of an open rectangle is free. The set τ is then the homotheties which transform the free boundary into a prescribed horizontal straight line. Thus the method is the simplest of the kind and it works quite well except perhaps when the unknown solution is too far off the prescribed horizontal; in this case the finite elements grid becomes rather peculiar so that the discretized problem is far from the continuous one (see figure 3).

Morice (1974) worked also in these lines but used

$$\tau = \{\text{conformal mappings of } \Omega_S \text{ into } \Omega_0\}\ ^{(*)}$$

The advantage of this method is that the formulation (4.6) is much simpler and there are ways to keep the discretization grid uniform

---

*Morice later extended this idea to more general implicit mappings (quasi-conformal mappings).

during the iterations (see figure 4). Thus at the cost of increasing the number of state variables, the method is very good. Unfortunately it is not completely general. Indeed if Ω_S is not simply connected it becomes very difficult to find a conformal mapping for it.

To conclude this paragraph, let me say that a compromise between the "easy to find" τ's (best would be an implicit method) and the τ's that makes (4.6) simple is still to be found.

4.2. <u>A descent method in the variable domain: the continuous case</u>

Suppose that $S = \{x(s)|s\in[0,1]\}$ is a solution of (4.4) and let $S' = \{x(s)+\lambda\alpha(s)n(s)|s\in[0,1]\}$ be an admissible boundary close to S (i.e. λ small) where $n(s)$ is the outward normal of S at $x(s)$. By definition if

(4.7) $$E(S) = \int_{\Omega_S} |\phi^S-\phi_d|^2 dx \qquad \text{(with self explanatory notations)}$$

then $$E(S') - E(S) \geq 0 \quad \forall S' \in \varepsilon$$

Let δE be the left member of (4.7). Suppose that $\alpha > 0$, by (4.7)

$$\frac{1}{2}\,\delta E = \int_{\Omega_S} [|\phi^S-\phi_d|^2-|\phi^{S'}-\phi_d|^2]dx + \int_{\Omega_{S'}-\Omega_S} |\phi^{S'}-\phi_d|^2 dx$$

it can be shown that $\|(\phi^{S'}-\phi^S)\|_{L^2} \to 0$ when $\lambda \to 0$, so that, by differentiation and from the mean value theorem (since $\Omega_{S'}-\Omega_S$ is a narrow strip around S)

$$\frac{1}{2}\,\delta E = \int_{\Omega_S} (\phi^S-\phi_d)\delta\phi dx + \int_S \lambda\alpha|\phi^S-\phi_d|^2 ds + o(\lambda)$$

where $\delta\phi = \phi^{S'}-\phi^S$. The P.D.E. in (4.4) in variational form is

(4.8) $$\int_{\Omega_S} (\nabla\phi\nabla w+\phi w)dx = \int_{\Omega_S} fw dx \quad \forall w \in H^1(\Omega_S)\ .$$

Therefore $\delta\phi$ satisfies

$$(4.9) \qquad \int_{\Omega_S} (\nabla\delta\phi\nabla w+\delta\phi w)\,dx = \int_S \lambda\alpha(fw-\nabla\phi\nabla w-\phi w)\,dS \ .$$

Thus if we let ψ be the solution of

$$(4.10) \qquad \int_{\Omega_S} (\nabla\psi\nabla w+\psi w)\,dx = \int_{\Omega_S} (\phi^S-\phi_d)w\,dx \quad \forall w \in H^1(\Omega_S)$$

Putting $w = \psi$ in (4.9) and $w = \delta\phi$ in (4.10) leads to

$$\int_{\Omega_S} (\phi^S-\phi_d)\delta\phi\,dx = \int_S \lambda\alpha(f\psi-\nabla\phi\nabla\psi-\phi\psi)\,dS$$

so that

$$(4.11) \qquad \frac{1}{2}\,\delta E = \int_S \lambda\alpha(|\phi^S-\phi_d|^2 + f\psi-\phi\psi-\nabla\phi\nabla\psi)\,dS + o(\lambda)$$

A similar computation can be done for (4.5) if one notes that in Ω_S $\delta\phi$ is a solution of $\Delta\delta\phi = 0$, $\delta\phi|_S = -\lambda\alpha\frac{\partial\phi^S}{\partial n} + o(\lambda)$ so that

$$(4.12) \qquad \delta E = 2\int_S \lambda\alpha(|\phi^S-\phi_d|^2 - \frac{\partial\psi}{\partial n}\frac{\partial\phi}{\partial n})\,dS + o(\lambda)$$

where ψ is the solution of

$$(4.13) \qquad \Delta\psi = \phi^S-\phi_d \ \text{ in } \Omega_S, \ \psi|_S = 0 \ .$$

It can be shown that (4.11) and (4.12) are valid for non positive α's.

Therefore let us consider the following algorithm for solving say (4.4) where $\mathcal{S}$ = {S twice differentiable} .

Algorithm 1

Step 0 Choose $S_0 \in \mathcal{S}$ set $i = 0$

Step 1 Compute $\phi^i = \phi^{S_i}$ by solving (4.8) with $S = S_i$

Step 2 Compute ψ^i by solving (4.10) with $S = S_i$, $\phi^S = \phi^i$

Step 3 Set $\alpha^i = -(\phi^i - \phi_d)^2 - f\psi^i + \phi^i\psi^i + \nabla\phi^i\nabla\psi^i$

Step 4 Compute λ_i solution of

$$\min E(S'(\lambda)) \text{ with } S'(\lambda)=\{x_{i+1}(s)=x_i(s)+\lambda\alpha_i(s)n_i(s)\,|\,s\in[0,1]\} . \tag{4.14}$$

and set $S_{i+1} = S'(\lambda_i)$, $i = i + 1$ and go back to 1 .

THEOREM 1

Any C^2-accumulation point of a sequence $\{S_i\}_{i\geq 0}$ generated by algorithm 1 is a stationary point of (4.3) and satisfies

$$|\phi^S - \phi_d|^2 + (f-\phi)\psi - \nabla\phi\nabla\psi = 0 \tag{4.15}$$

Proof: Algorithm 1 is a method of steepest descent for (4.4) locally in the space of admissible α's such that $S' \in \mathcal{S}$. The convergence proof proceeds just as in the case of an ordinary gradient method. From (4.11) with α as in step 3 ,

$$E_{i+1} - E_i = -\lambda_i \int_{S_i} [|\phi^{S'} - \phi_d|^2 + (f-\phi^i)\psi^i - \nabla\phi^i\nabla\psi^i]^2 dS_i + o(\lambda_i)$$

So that each iteration decreases E of a non trivial quantity until the quantity in (4.15) is zero. For a rigorous proof see Pironneau (1976).

4.3. Implementation of algorithm 1

Thus algorithm 1 proceeds like the method of relaxation except that its convergence can be proved, at the cost of integrating a second P.D.E., the adjoint system.

There are cases where algorithm 1 can be implemented directly:
i) when the numerical integration of the PDE's can be done with a very good precision (see figure 5);
ii) when the cost function of the optimum design problem is very sensitive to the positions of the free boundary.

In all other cases a simple discretization of the equations of continuous problem would fail to enable us to solve the optimum design problem with a good precision because the numerical noise of the discretization makes it impossible to find a positive λ solution of (4.14).

Therefore let us derive a formula similar to (4.11) for the discrete case.

For a given S_o, let τ_h be a triangulation of Ω_h, a polygonal approximation of Ω_{So},

$$\text{i.e. } \Omega_h = \bigsqcup_{i=1}^{n} T_h^i \ , \ T_h^i \cap T_h^j = \text{one side or one node,}$$

T_h^i = triangle where h denotes the length of the largest side, let

$$H = \{w | w \text{ linear on } T_h^i\}, \ i=1,\ldots,n \ ; \ w \text{ continuous}$$

If w_j is the function in H which equals 1 at the node j and zero at all other nodes, then $\{w_j\}_{j=1}^m$ is a basis for H (m = number of nodes).

A first order finite element discretization (4.8) is (r = number of interior nodes)

$$\text{(4.16)} \qquad \int_{\Omega_h} (\nabla\phi\nabla w_j + \phi w_j)dx = \int_{\Omega_h} f w_j dx \quad j=1,\ldots,r \ ; \ \phi = \sum_{j=1}^{r} \phi^j w_j \ .$$

Equation (4.16) is a linear system in ϕ^j which can be solved by a relaxation method, for example.

Similarly (4.10) is approximated by

$$\text{(4.17)} \qquad \int_{\Omega_h} (\nabla\psi\nabla w_j + \psi w_j)dx = \int_{\Omega_h} (\phi-\phi_d)w_j dx, \ j=1,\ldots,m \ ; \ \psi = \sum_{j=1}^{m} \psi^j w_j$$

where ϕ is the solution of (4.16)

Let $S'(\lambda)$ be the boundary obtained from S_0 by moving the nodes along the lines of discretization "perpendicular" to Ω_h (see figure 6) then the discrete analogue of (4.11) can be obtained in a similar way:

$$\frac{1}{2}\,\delta E_h = \int_{\Omega'_h - \Omega_h} (|\phi-\phi_d|^2 + f\psi - \phi\psi - \nabla\phi\nabla\psi)\,dx + o_h(\lambda) \tag{4.18}$$

In establishing (4.18) we have used the fact that ϕ and ψ depend continuously upon the coordinates of the nodes. Now from the mean value formula for integrals

$$\frac{1}{2}\,\delta E_h = \int_{S_0} (|\phi-\phi_d|^2 + f\psi - \phi\psi - \nabla\phi\nabla\psi)\,\lambda\alpha dx + o_h(\lambda) \tag{4.19}$$

Thus the discrete version of algorithm 1 is

Algorithm 2: We assume that the user has an automatic triangulation subroutines which generate the interior nodes from the boundary nodes. For simplicity we assume that the first point of S_0 remains fixed.

Step 0 - Choose S_0 , set $i=0$, choose τ_h , choose $\beta \in (0,1)$

Step 1 - Compute ϕ from (4.16) with $\Omega_h = \Omega_h^i$

Step 2 - Compute ψ from (4.17) with $\Omega_h = \Omega_h^i$

Step 3 - Let ℓ be the number of boundary nodes on S , set

$$\alpha^i(s) = \alpha_j^i + \frac{s-s_j}{s_{j+1}-s_j}(\alpha_{j+1}^i - \alpha_j^i) \text{ for } s \in (s_j, s_{j+1}),\ j=1,\dots,\ell \tag{4.20}$$

where s_j is the curvilinear abcissa of the node j and α_j^i are determined as follows: let σ_i be the sines of the angle of the discretization lines, let

$$\delta E_{hj}^i = \int_{S_i} (|\phi-\phi_d|^2 + (f-\psi)\phi - \frac{\partial\phi}{\partial s}\frac{\partial\psi}{\partial s})\,\alpha^{ij}(s)\,\sigma_i ds \tag{4.21}$$

where $\alpha^{ij}(s)$ is given by (4.20) with $\alpha_k^i = \delta_{jk}$, $k=1,\dots,\ell$

(4.22) then take $\alpha_j^i = -\delta E_{hj}^i/(s_j-s_{j-1})$ $j=2,\dots,m$; $\alpha_1^i = 0$

Step 4 - Obtain $S(\lambda)$ by moving the nodes of S_i on the discretization lines of a quantity $\lambda\,\alpha_j^i$ and compute the smallest integer k such that (with self explanatory notations)

$$(4.23)\qquad E_h(S(\lambda)) - E_h(S_i) \le -\beta\lambda \sum_{j=1}^{\ell} (\alpha_j^i)^2(s_j-s_{j-1}),\ \lambda = (\tfrac{1}{2})^k$$

and such that $S(2^{-k})$ is not nearer than, say $\frac{1}{3}$ of the distance of S_i , to the next discretization line "parallèle" to S_i .

Step 5 - Set $S_{i+1} = S(2^{-k})$ compute a new triangulation around S_{i+1} from its nodes computed in step 4, set $i = i+1$ and go back to step 1.

Conjecture - If the automatic triangulation subroutine is such that the position of the interior nodes depend continuously upon the position of the boundary nodes (and it is the case of some such subroutines, for a given ℓ) then the sequence $\{E_h(S_i)\}_{i>0}$ will be a decreasing sequence and $\{S_i\}$ will converge to an accumulation point of the discrete analogue of (4.4).

Justification - First of all (4.23) has always a non zero solution because

$$-\sum_{j=1}^{\ell} (\alpha_j^i)^2(s_j-s_{j-1})$$

is the slope of $\lambda \to E_h(S(\lambda))$ at $\lambda = 0$ (see (4.19)(4.21)(4.22)). Therefore algorithm 2 is the method of steepest descent applied to (4.4) with $\Omega = \Omega_h$ and (4.16) instead of the continuous PDE; with one difference however, the triangulation is rebuilt at each iteration (so that each triangulation is well suited to the new geometry). Therefore, since the cost function changes with the discretization, $E_h(S_{i+1})$ with the new triangulation might be greater than $E_h(S_i)$, despite (4.23). It should be noted that it might be possible to prove the convergence of algorithm 2 by using a model similar with the one described by Klessig-Polak (1972), by asking to increase the number of nodes when the cost increases. However, the following argument seems to indicate that the cost is unlikely to be increasing.

It is theoretically possible to consider E_h as a function of the coordinates of all the nodes (instead of the boundary nodes only). And indeed in doing so one is very close to the concept of mappings of Ω_S into Ω_0, studied in the previous section. O'Carroll et al (1976) have tested this method, by the way. Now the change of cost due to a shift of a boundary node tangentially to S is certainly smaller than the change of cost due to a normal shift. In the first case one changes the definition of the continuous problem also while in the second case one changes the discretization of the problem. It seems also that the change of cost due to a shift of the interior nodes would be of the same order, if not smaller, than for the above tangential shift.

Similarly, an implementable algorithm can be obtained for problem (4.5). To derive an analogue of (4.12) one needs to use the following formula (Murat-Simon (1974)): if ϕ satisfies

$$\int_\Omega \nabla\phi\nabla w dx = \int_\Omega f w dx \quad \forall w \in H^1_o(\Omega) \; , \phi \in H^1_o(\Omega)$$

then when Ω is replaced by Ω', $\delta\phi \in H^1_o(\Omega)$ satisfies

$$\int_\Omega \nabla\delta\phi\nabla w dx = -\int_\Omega \lambda\nabla w\cdot(f\vec{\theta} + \operatorname{div}\theta\vec{\nabla}\phi - (\theta' + {}^t\theta')\nabla\phi)dx + o(\lambda) \quad \forall w \in H^1_o(\Omega)$$

$(x,y) + \lambda\theta(x,y) = (x+\lambda\,\theta_1(x,y),\, y+\lambda\,\theta_2(x,y))$ are the coordinates of the transform of (x,y) by the map $I+\lambda\theta$ that transforms Ω into Ω'. Then the analogue of (4.19) is

$$\delta E_h = -2\int_\Omega \lambda\nabla\psi[f\theta + \operatorname{div}\theta\nabla\phi - (\theta' + {}^t\theta')\nabla\phi]dx + \int_S \lambda\alpha|\phi^S - \phi_d|^2 dS + o_h(\lambda)$$

where ψ is the solution in $H^1_o(\Omega)$ of

$$\int_\Omega \nabla\psi\nabla w dx = \int_\Omega (\phi^S - \phi_d) w dx \quad \forall w \in H^1_o(\Omega)$$

These equations remain valid when Ω is changed into Ω_h and H^1_o into κ. The reader will easily construct an implementation of algorithm 1 in this case, by letting $\theta=0$ except in the strip of triangles that touches S, where θ is taken linear and in tune with α.

5. CONCLUSION

An engineer faced with a free boundary problem may be in one of the two positions: either he knows very well his problem intuitively and he is able to find a good estimate of the solution; or he does not know much about the answer and he is not willing to develop a special subroutine to find it. Then, if he is in the first case he will be probably better off by using the relaxation method; but if he is in the second position he would prefer to call an "automatic" subprogram. Such efficient automatic scheme is still to be found; the last section of this paper contains one that looks reasonably promising on account of the fact that free boundary and optimum design problems are numerically usually quite stiff.

ACKNOWLEDGMENT

I wish to thank J. R. Bourgat, A. Dervieux and P. Morice for their helpful suggestions.

REFERENCES

1. Baiocchi C. (1975) - Cours au Collège de France.
2. Begis D., Glowinski R. (1975) - Application de la méthode des éléments finis à l'approximation d'un problème de domaine optimal. Applied Math. & Optimization, Vol. 2, N° 2.
3. Benssoussan A., Lions J. L. (1973) - CRAS 276, pp. 1189-1193, Paris.
4. Bourgat J. F., Duvaut G. (1975) - Calcul numérique avec ou sans sillage autour d'un profil bidimensionnel symétrique et sans incidence. Rapport Laboria N° 145.
5. Bourgat J. F., Duvaut G. (1976) - Résolution numérique d'un problème de Stephan à 2 phases par une inéquation variationnelle (à paraître).
6. Bourot J. (1974) - CRAS A278.455.
7. Céa J. (1975) - Une méthode numérique pour la recherche d'un domaine optimal. Proceedings I.F.I.P. Congrès Nice.
8. Céa J., Glowinski R. (1972) - Méthodes numériques pour l'écoulement laminaire d'un fluide rigide visco-plastique incompressible. Intern. J. of Computer, Math. B, Vol. 3, pp. 225-255.
9. Chavent G. (1971) - Thèse de Doctorat, Paris.
10. Chesnais D. (1975) - On the existence of a solution in a domain identification problem. J. of Math. Anal. and Appli. 52,2.
11. Cryer C. W. (1970) - On the approximate solution of free boundary problems using finite differences. J. of the Association for Comp. Machinery, Vol. 17, N° 3, pp. 397-411.
12. Garabedian P. R. (1956) - The mathematical theory of three dimensional cavities and jets.Bull. Amer. Math. Soc. 62, pp. 219-235.

13. Garabedian P. R. (1964) - Partial Differential Equations. Wiley, New York.
14. Glowinski R., Lions J. L., Tremoliere R. (1976) - Analyse numérique des inéquations quasi-variationnelles. Dunod (to be published).
15. Hadamard J. (1910) - Leçons sur le calcul des variations. Gauthiers-Villars.
16. Klessig R.,Polak E. (1970) - A method of feasible directions using function approximation with applications to min-max problems. E.R.L.-M287, University of California, Berkeley.
17. Lamb H. (1879) - Hydrodynamics. Cambridge University Press.
18. Lions J. L. (1972) - Some aspects of the optimal control of distributed parameter systems. RESAM, 6 SIAM.
19. Murat R., Simon J. (1974) - Quelques résultats sur le contrôle par un domaine géométrique. Université de PARIS VI, Rapport interne Lab. Analyse Numérique, N° 74003.
20. Morice P. (1974) - Une méthode d'optimisation de forme de domaine. Proc. Congrès IFIP-IRIA, Paris. Springer Verlag.
21. O'Carroll M. J. and H. T. Harrison (1976) - Proc. ICCAD Conf. Rappalo, Italy.
22. Pironneau O. (1973) - On optimum profiles in Stokes flow, J. Fluid Mech., Vol. 59, pp. 117-128.
23. Pironneau O. (1976) - Thèse de Doctorat, Paris.
24. Southwell R. V. (1946) - Relaxation methods in theoretical physics, Vol. 1, Clarendon Press, Oxford.
25. Stanitz J. (1952) - Design of two-dimensional channels, NACA technical note 2595.
26. Trefftz E. (1916) - Uber die Kontraktion kreisförmige Flüssigkeitsstrahlen. Z. Math. Phys. 64, pp. 34-61.
27. Zienkiewicz O. C. (1971) - The finite elements in Engineering Science. McGraw-Hill, London.

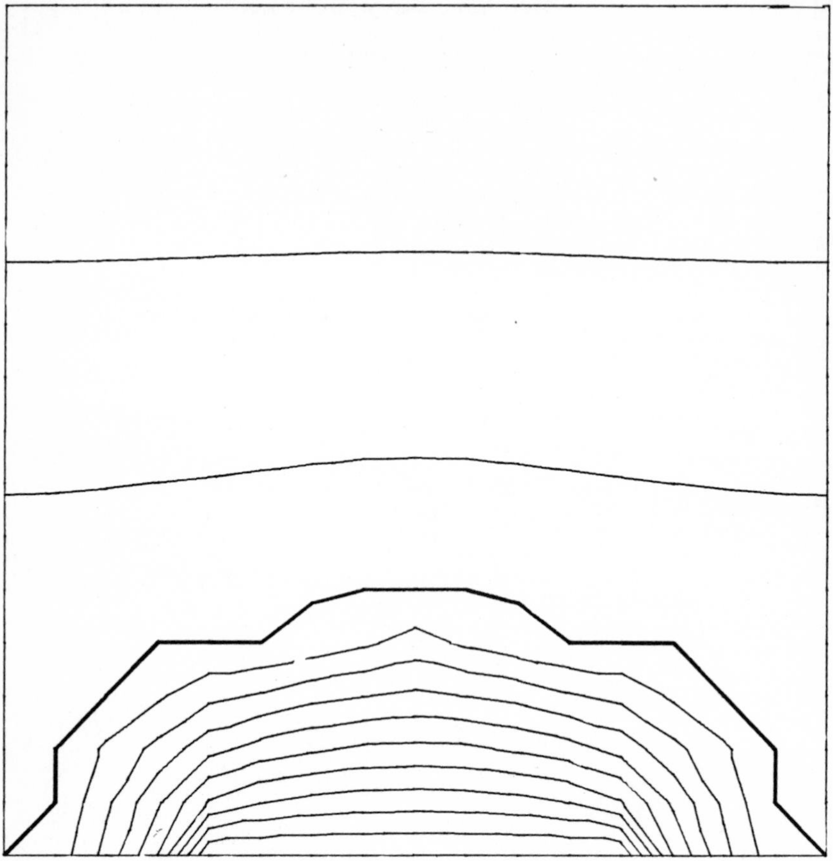

Figure 1 - (From Bourgat-Duvaut (1976)).
Fusion of an ice cube when the bottom plane is heated. Even though the free boundary is not well approximated, the isothermal lines are computed accurately.

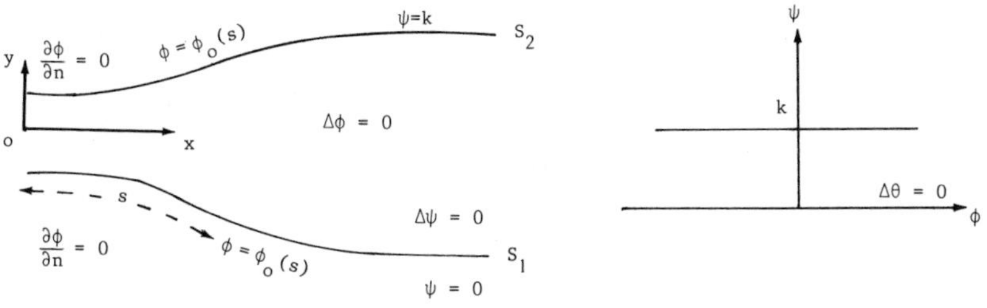

Figure 2

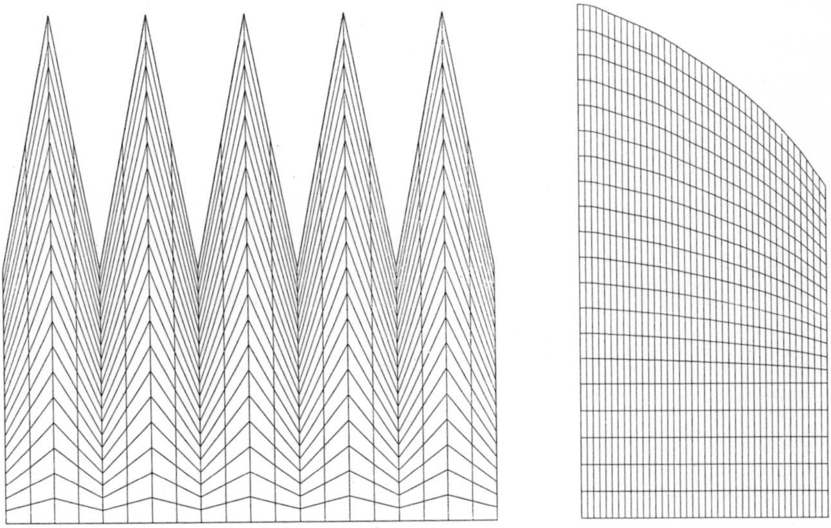

Figure 3 - (From Begis-Glowinski (1975)).

Computation of a free boundary (upper line) by the method of Begis-Glowinski and the corresponding finite element triangulation in a good case and in a bad case.

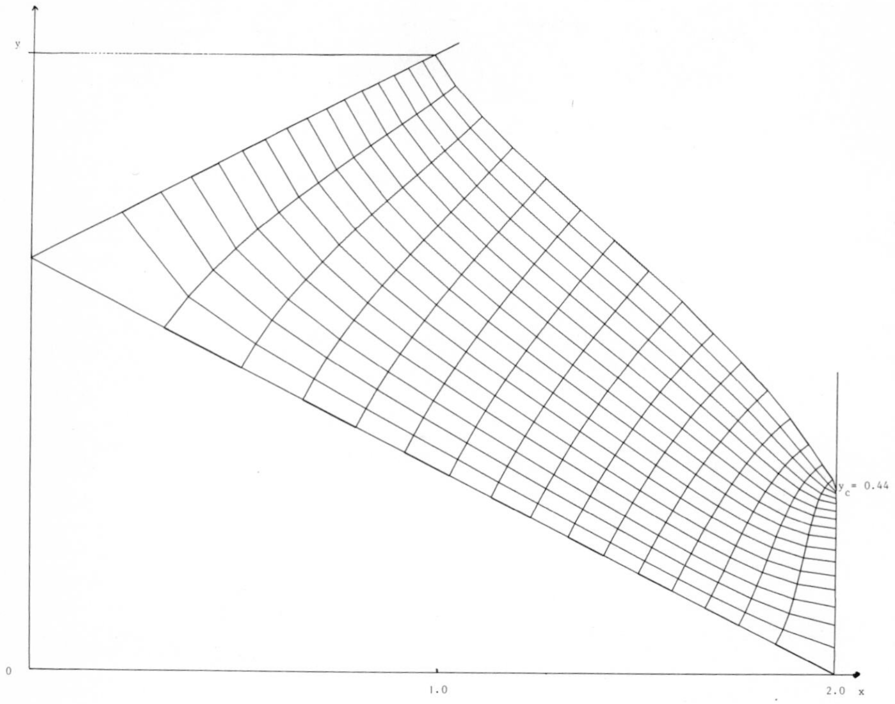

Figure 4 - (From Morice (1974)).
Computation of a free boundary by Morice's method and the corresponding triangulation.

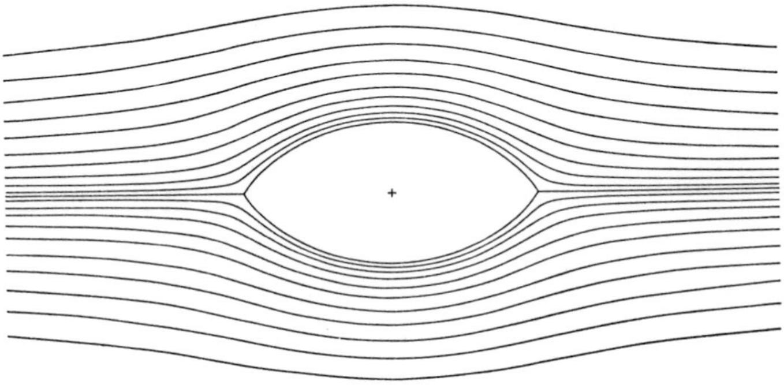

Figure 5 - (From Bourot (1974)).

Computation of the minimum drag profile in Stokes flow by the method of Pironneau (1973).

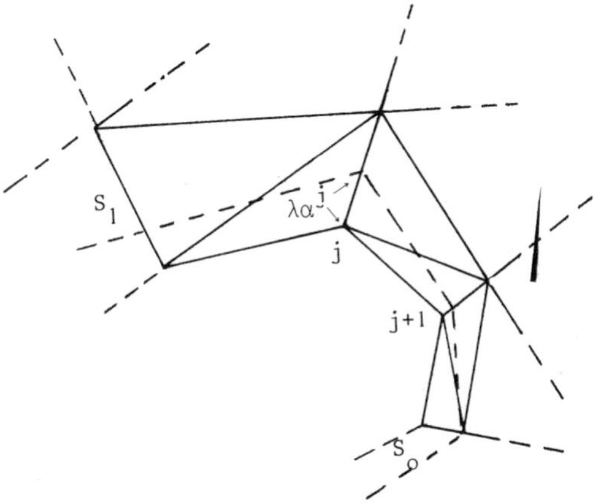

Figure 6

S_1 is obtained from S_0 by moving the nodes of discretization of S_0 of a suitable quantity: $\lambda\alpha^j$, along the lines of discretization of ΩS_0 "perpendicular" to S_0 .

"SOME APPLICATIONS OF STATE ESTIMATION AND CONTROL THEORY TO DISTRIBUTED PARAMETER SYSTEMS"

W. H. Ray

I. Introduction

In this chapter we shall discuss a large number of real and potential applications of state estimation and control theory to distributed parameter systems. Because of the inherent practical difficulties of state measurement in distributed systems, the major problems in control implementation are often due to difficulties in obtaining reliable estimates of the state variables. For this reason, the major emphasis of this chapter shall be on distributed parameter state estimation. The reader is referred to other survey articles for an overview of parameter identification [1,2] and to the other chapters in this volume for a discussion of the control of distributed parameter systems. A second troublesome problem arising in applications is the development of efficient computational algorithms which will allow real time implementation of these state estimation and feedback control schemes. In what follows, we shall give examples of how these numerical problems can be solved in practice.

The next section gives an overview of practical considerations and the algorithms available for distributed parameter state estimation and presents a survey of reported applications. Following this, a detailed case study of the real time implementation of state estimation and feedback control to a laboratory model process will be discussed in some detail. Finally, a general survey of remaining practical problems will be presented along with some suggestions for future research directions.

II. A Survey of State Estimation Algorithms and Applications

The field of sequential state estimation has grown enormously in scope and popularity since the original work of Kalman [3] and Kalman and Bucy [4] in the early 1960's. However, it was not until 1967 that the first extensions of these ideas to distributed parameter systems began to appear. The earliest paper seems to be due to Kwakernaak [5], and was quickly followed by the work of Falb [6], Tzafestas and Nightingale [7-10], Balakrishnan and Lions [11], Thau [12], and Seinfeld [13], all of whom apparently developed their results independently and nearly simultaneously. Since these early papers, work has progressed so that state estimators are now available for numerous classes of linear and non-linear distributed parameter systems.

Unfortunately, the reported practical applications of distributed parameter estimation algorithms has badly lagged the theory. This is perhaps due to the relative complexity of the estimation equations when compared to the classical Kalman-Bucy filter. Another factor is the lack of meaningful contact between potential industrial users and research engineers familiar with the capabilities of these techniques.

In this section we shall summarize the results available for distributed parameter estimation and describe recent computational and experimental example applications. In this way it is hoped to provide the potential user of these techniques with a guide to the theory as well as inspiration for further applications.

This survey section, which has been updated from an earlier lecture [14] by the author, shall begin with a discussion of the theoretical results available for both linear and nonlinear systems and then review the example applications which have appeared to date.

A. Theoretical Results

There have been a large number of theoretical approaches taken in the derivation of distributed parameter state estimators. The most direct approach is to somehow approximately lump the distributed system and then apply the classical lumped parameter theoretical results [3,4]. One can certainly proceed in this way and obtain useful results (e.g., [15-20]), although in certain cases the distributed parameter formulation does seem to have advantages (e.g., when there is boundary noise). Because there are so many ways of lumping a distributed system (e.g., finite difference, modal representation, splines, etc.) it is hard to compare exact distributed parameter results with lumped approximations, and thus the possible pitfalls of lumping have not been well defined. In the discussion of theoretical results which follows, we shall deal mainly with results derived for the full distributed parameter system model; however, several cases of "lumping approaches" to filtering will be included in the applications section.

Even excluding lumping, there have been a wide range of approaches taken in arriving at theoretical results ranging from "formal" techniques such as minimizing a given error functional to more "rigorous" approaches explicitly treating the evolution of the probability distribution of the state variables. While the more rigorous approaches yield more precise statistical information about the problem, they are limited in the classes of problems which may be treated. The formal approaches yield less detailed statistical information, but have the ability to treat a wide range of both linear and non-linear problems. Therefore, both methods of analysis have value and in fact are usually found to yield identical filter equations in situations where both techniques apply.

Because this survey section is emphasizing applications, it is not our intention to attempt an extensive review of the theoretical approaches for developing distributed parameter state estimators, but rather shall concentrate on the results which may be of direct interest to the practitioner. For a fuller discussion of the theoretical ideas see the review by Tzafestas [21].

1. Observability and Measurement Location

Perhaps the first consideration in the application of a state estimator is the question of observability and the optimal location of measurement devices. Observability, which has been defined for lumped parameter systems as the ability to determine the initial state based on the system model and measurements available, is a more difficult question for distributed parameter systems because of the strong influence of the location of the sensing devices. Thus for a fixed number of sensors, it is possible to choose locations for which (i) the system is not observable, or (ii) the system is observable, or (iii) for which the system is both observable and the measurement locations yield the maximum amount of information in the presence of noise. For these reasons, the choice of type, number, and location of sensors is crucial in the design of a distributed parameter state estimator.

Wang [22] seems to be the first to discuss observability for distributed parameter systems and proposed a definition based on the recovery of the initial state through analogy to lumped parameter systems. Goodson and Klein [23] define observability as the existence of a unique system state based on the model and the observation device and have developed general criteria for observability for a number of linear partial differential equation systems. For first order quasi-linear systems, their criteria includes the requirement that each of the characteristic lines in the system intersect a sensing device. Yu and Seinfeld [24] have quantified this requirement and, by applying the known lumped parameter observability theorems along the characteristic lines of the system, have developed criteria specifying which combinations of states must be measured. Their results take the familiar form of determining the rank of an observability matrix. Recently Thowsen and Perkins [25] have extended these results to a slightly more general form of the equations, and Sendaula [26] has considered the related problem of time delay systems.

Goodson and Klein [23] also consider linear diffusion and wave equations and are able to develop sufficient conditions for observability based on the choice of sensor placement. Cannon and Klein [27] have numerically investigated the optimal placement of the measurement location for a simple linear example problem of heat conduction with specified surface temperatures. Yu and Seinfeld [28] have developed sufficient conditions for observability for linear systems amenable to a modal representation. Their approach is to represent the system in terms of N eigenfunctions and then apply the known observability conditions [3,4] for lumped parameter systems to the resulting ordinary differential coefficient equations. The rank of the observability matrix depends on the values of the eigenfunctions at the measurement locations. These results were applied to both one and two dimensional heat conduction and diffusion examples. In addition, numerical studies were performed to show the influence of the measurement location on the quality of the estimates.

In spite of these valuable contributions, at present there exists no general theory of observability for either linear or nonlinear distributed parameter systems; there exists only a few results for the particular linear cases studies. Nevertheless, several recent papers [29,30] would seem to lead toward such a theory. Similarly no general results are available for determining optimal sensor locations; all that one can do is numerically investigate each problem individually for a range of model parameters, noise statistics, etc. Several workers have developed systematic algorithms for carrying this out [31-33]. However, further theoretical results would be extremely useful here.

2. State Estimators Available

State estimators, for smoothing, filtering, and prediction, have been developed for a wide range of distributed parameter systems. In this section we shall discuss explicitly the filtering results; however, the related prediction and smoothing algorithms can be found in a straightforward manner once the filtering equations are known. Just as in the case of lumped parameter systems, linear distributed parameter systems allow the possibility of much more rigorous and precise results than do nonlinear systems. Not only do the stochastic system modelling equations have precise meaning, but the estimate covariances arise naturally in the filter calculations and provide valuable statistical information about the quality of the estimates. On the other hand nonlinear state estimators, developed by more formal techniques, are much more general, greatly expand the range of practical problems which can be handled, and have shown to reduce to known, rigorously derived, linear filters when the nonlinear equations are made linear. Thus the practitioner is advised to take advantage of the rigorous linear filter, with its precise estimate statistics, where possible, and to make use of the more general nonlinear filtering results, when necessary. Because of the fact that the independently derived linear filters appear as special cases of the more general nonlinear results, in what follows we shall group the state estimation results under their general nonlinear form. A summary of available theoretical results is presented in Table 1.

The first class of distributed parameter systems to have filters developed were systems with time delays. Kwakernaak's classic paper [5] led the way and showed the structure of the distributed parameter filtering problem. Although a number of linear filtering results have appeared since, [17,34-42], for the purpose of our discussion here these can be treated as special cases of the nonlinear filtering results of Yu et al. [43]. In this general formulation, the filtering problem for time delays takes the form

$$\dot{\underset{\sim}{x}} = \underset{\sim}{f}(\underset{\sim}{x}(t), \underset{\sim}{z}(r_1,t), \;--\underset{\sim}{z}(r_\beta,t)) + \int_0^1 \underset{\sim}{K}(\underset{\sim}{z}(r,t)dr + \underset{\sim}{\xi}(t) \tag{1}$$

$$\underset{\sim}{z}_t(r,t) = -\underset{\sim}{M}(r,t)\underset{\sim}{z}_r(r,t) + \underset{\sim}{g}(\underset{\sim}{z}(r,t),r,t) + \underset{\sim}{\zeta}(r,t) \tag{2}$$

$$\underset{\sim}{y}(t) = \underset{\sim}{h}(\underset{\sim}{x}(t), \underset{\sim}{z}(r_1{*},t), \;--\underset{\sim}{z}(r_\gamma{*},t),t) + \int_0^1 \underset{\sim}{H}(\underset{\sim}{z}(r,t),r,t)dr + \underset{\sim}{\eta}(t) \tag{3}$$

$$\underset{\sim}{x}(0) = \underset{\sim}{x}_0 \tag{4}$$

$$\underset{\sim}{z}(r,0) = \underset{\sim}{z}_0(r) \tag{5}$$

$$\underset{\sim}{z}(0,t) = \underset{\sim}{b}(\underset{\sim}{x}(t)) \tag{6}$$

Table 1. Summary of Theoretical Results

Topic	References
Systems with time delays	[5,17,34-43,89]
Systems described by second order partial differential equations	Linear Systems [7-9,11,12,27, 28,45-64] Nonlinear Systems [10,13,15, 44,58,65-67]
Linear systems described by integral or integro-differential equations	[6,9,47,48,54,55,57,58]
Systems described by partial differential equations with moving boundaries	[68]
Separation principles and stochastic feedback control	[8,35,38,52,57,69,70]

where Eqs. (1,2) represent the behavior of a coupled lumped and distributed parameter system defined for $t \geq 0$ and on the spatial domain $r \in [0,1]$. $\underset{\sim}{x}(t)$ and $\underset{\sim}{z}(r,t)$ are state vectors with boundary conditions (4-6). Observations consist of the output vector $y(t)$ defined by (3). The quantities $\underset{\sim}{\xi}(t)$, $\underset{\sim}{\zeta}(r,t)$, and $\underset{\sim}{\eta}(t)$ represent zero mean random processes with arbitrary statistical properties. Within this framework an extremely wide range of time delay problems may be treated.

A second class of distributed parameter systems for which a large number of results are available are second order partial differential equations systems. A very general form of this problem for one spatial variable is [44]

$$\underset{\sim}{z}_t(r,t) = \underset{\sim}{g}(r,t,\underset{\sim}{z},\underset{\sim}{z}_r,\underset{\sim}{z}_{rr}) + \underset{\sim}{\xi}(r,t) \qquad t > 0,\ 0 \le r \le 1 \tag{7}$$

$$\underset{\sim}{b}_0(t,\underset{\sim}{z},\underset{\sim}{z}_r) + \underset{\sim}{\gamma}_0(t) = 0 \qquad r = 0 \tag{8}$$

$$\underset{\sim}{b}_1(t,\underset{\sim}{z},\underset{\sim}{z}_r) + \underset{\sim}{\gamma}_1(t) = 0 \qquad r = 1 \tag{9}$$

$$\underset{\sim}{y}(r,t) = \underset{\sim}{h}(r,t,\underset{\sim}{z}(r,t)) + \underset{\sim}{\eta}(r,t) \tag{10}$$

where $\underset{\sim}{z}(r,t)$ is a vector state variable modeled by (7-9) and $\underset{\sim}{y}(r,t)$ is a vector of measurements defined by (10). $\underset{\sim}{\xi}(r,t)$, $\underset{\sim}{\gamma}_0(t)$, $\underset{\sim}{\gamma}_1(t)$ are zero mean volume and boundary noise with arbitrary statistical properties and $\underset{\sim}{\eta}(r,t)$ is a zero mean measurement error. A large number of powerful linear results have appeared for special cases of this class of problems [7-9,11,12,27,28,45-64], and a number of less rigorous but fairly general non-linear estimators have been developed as well [10,13,15,44,58,65,66]. However, for purposes of classification all of the resulting filter equations can be considered special cases of the filter developed by Hwang et al. [44] for the general nonlinear formulation (7-10). Recently, Ajinkya et al. [67] have extended these nonlinear results to include systems described by coupled ordinary and second order partial differential equations.

A number of linear results have also been derived for systems described by integral or integro-differential equations [6,9,47,48,54,55,57,58], but no explicit examples seem to have been discussed.

Very recently a general nonlinear state estimator for systems described by second order partial differential equations and having moving boundaries was developed [68]. As will be discussed later, this filter allows the simultaneous estimation of the system state as well as the boundary position for a large number of problems of practical interest.

It should also be noted that some work has appeared on separation principles and the feedback control of some types of linear stochastic systems [8,35,38,52,57,69,70]. In addition, there seems to be a rapidly growing literature on distributed parameter observers [42,71-74].

From the papers summarized here, it is clear that there are state estimators available for many classes of distributed parameter systems. In the next section we shall discuss a number of practical engineering problems for which these theoretical results are applicable.

B. AN OVERVIEW OF APPLICATIONS

Before beginning a detailed discussion of particular applications of distributed parameter state estimators, it is useful to consider some practical points in the implementation of such filters. Firstly, let us recall that the great majority of filtering algorithms discussed in the previous section assume measurements continuous in time and often even in space. Certainly this is not the case in practice where measurements are usually taken at specific spatial locations and quite often at fixed discrete intervals of time. It is fortunate that this apparent inconsistency between the available theory and industrial practice is not a serious limitation - due to some simple tricks. As has been shown by Meditch [48] one may convert a continuous spatial measurement device estimator to a discrete spatial measurement result by making use of a Dirac delta function of the form $\delta(r-r_i^*)$ where $r_i^* i = 1,2,\text{--}$ are the actual measurement locations. This immediately produces the correct estimation equations for both linear [48] and nonlinear [44] systems with discrete spatial observations. A similar trick involving $\delta(t-t_k)$ may be used for data which must be taken at discrete time intervals, t_k [67]. However, in this case the approach is difficult to justify rigorously and thus must be used with care. Nevertheless, it has the advantage that it immediately converts continuous time estimation equations to discrete time measurement equations, and can be shown to produce the known rigorously derived discrete time filters (e.g.; [37]) from their continuous time analogs (e.g.; [5]).

A second practical consideration which should be discussed is the development of efficient computational algorithms for the on-line implementation of distributed parameter filters. Distributed parameter filtering algorithms require the solution of the distributed parameter modelling equations in real time. Furthermore the covariance equations turn out to be nonlinear partial differential equations in $2n + 1$ independent variables where n is the number of spatial variables in the model. Clearly the soltuion of such equations can quickly become unmanageable if some care is not taken in the choice of the computational algorithm. For linear systems, the problem is simplified because the covariance equations are independent of the data and system state and may be solved off-line. In addition, there are powerful analytical techniques which may be used for solving the state equations. For first order hyperbolic differential equations, one may use the method of characteristics to reduce the computations to solving ordinary differential equations along the characteristic lines [75,76]. For second order partial differential equations such as occur in heat, diffusion, or wave processes the solution can be represented in terms of the eigenfunctions of the system. This modal approach has been proposed by a number of workers [7,16,28,51,52,58,77,78] and reduces both the state and covariance equations to a set of ordinary differential equations. For nonlinear state estimation problems the computational problems are not so easily solved; however, this author feels that the use of weighted residual methods (such as Galerkin or Collocation procedures), which are the nonlinear analogs of eigenfunction expansions, should lead to filter calculations of manageable proportions.

In some applications it is the development of a reliable system model which limits the use of state estimation. For heat conduction problems this is seldom a difficulty, but for more complex processes such as chemical reactors, model development is usually the crucial overriding problem. Fundamentally, state estimation is a compromise between complete, errorless state measurement and perfect predictive modelling. Therefore, the less data available, the better the model must be and vice-versa. For these reasons, it is essential in practice that a detailed modelling study be carried out before state estimation begins.

The reported applications of distributed parameter state estimation can be classified as follows:

1. Computer simulations with computer generated data based on a model of more or less physical significance.

2. Off-line filtering of actual process data in which one "pretends" to filter in real time.
3. Actual real time state variable filtering and control of an operating process.

The great majority of the reported case studies are of type 1 [9,10,13,15,17-19,20,34,35,37,42-44,55,57,60,66,67,74,75,77,79-84], very few are of type 2 [85,86] and to the author's knowledge, there is only one paper [78] of type 3. With a few exceptions, most of the type 1 applications are example problems designed to demonstrate a new theoretical result rather than being feasibility studies of a genuine practical problem. The single reported experimental implementation [78] provides a great deal of insight into practical details of implementation as well as representing a real test of distributed parameter filtering and thus will be discussed in great detail later.

In an effort to provide application ideas to the practitioner, the existing literature of suggested, simulated, and experimental applications of distributed parameter state estimation will now be discussed within the framework of several practical state estimation problems. A summary of these problems with relevant references is given in Table 2.

1. Heat Conduction Processes

The first class of problems, the estimation of the temperature profile in a heat conductor has been used extensively as a demonstration problem for distributed filters [9,10,13,44,55,60,66,67,77,78,81]; however, this problem does arise in a number of industrial processes such as found in metallurgical operations [67], glass making [87], and nuclear reactor control [84,88]. As a particular example, let us consider the soaking pit operation in steelmaking. Ingots are heated in a furnace (soaking pit) preparatory to rolling into slabs. See Figure 1 for a sketch of the physical picture.

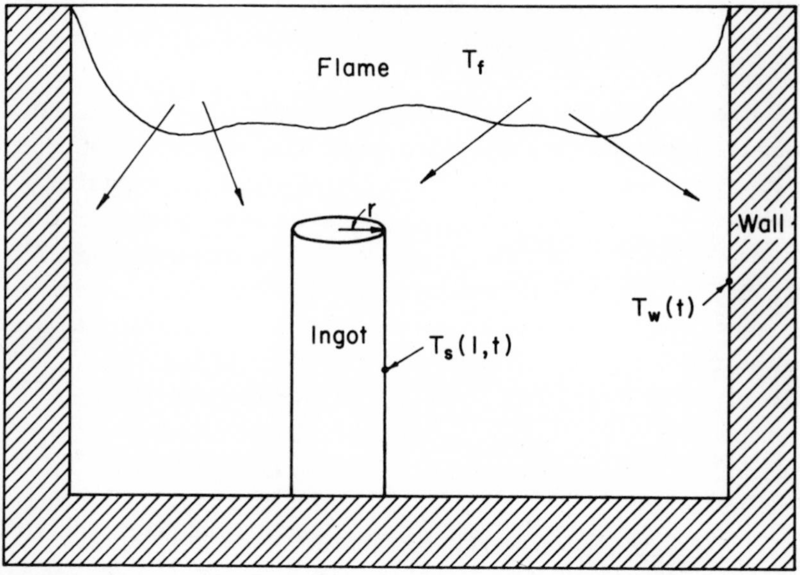

Figure 1. An ingot heated in a "soaking pit" furnace

Table 2. Summary of Example Applications

State Estimation Problem	References
Estimation of the temperature profile in a heat conductor	[9,10,13,44,55,60,66,67,77,78,81]
Estimation of temperature, concentrations, and catalyst activities in chemical reactors	[15,18-20,43,44,80,82,83,85,89]
Estimation of pollutant concentrations in air and water pollution monitoring problems	[75,76,86,92]
Estimation of states and boundary position in systems having moving boundaries	[68]

The problem is that the initial temperature distribution of the ingots when they go into the furnace is unknown and, at best, only noisy ingot surface termperature measurements are available. In spite of this, it is necessary to control the furnace heating rate and to remove the ingots from the furnace precisely when the temperature profile is satisfactory for rolling.

The modelling equations for a simple cylindrical geometry with no axial temperature variations are given by

$$\frac{\partial T_s}{\partial t} = \frac{\alpha}{R^2}\left\{\frac{\partial^2 T_s}{\partial r^2} + \frac{1}{r}\frac{\partial T_s}{\partial r}\right\} \qquad \begin{matrix} 0 \le r \le 1 \\ t \ge 0 \end{matrix} \tag{11}$$

which together with the boundary conditions

$$\frac{\partial T_s}{\partial r} = 0 \qquad r = 0 \tag{12}$$

$$\frac{\partial T_s}{\partial r} = a_s(T_f^{\,4} - T_s^{\,4}) + b_s(T_w^{\,4} - T_s^{\,4}) \qquad r = 1 \tag{13}$$

describe the temperature dynamics of the metal ingot. The variables T_f and T_w represent the flame temperature and the furnace wall temperature respectively. For this problem the ingot surface temperature, $T_s(1,t)$, can be measured intermittently (e.g.; by optical pyrometry), but with a large amount of measurement error. Thus the measurement device can be written

$$y(t_k) = T_s(1,t_k) + \eta(t_k) \qquad k = 1,2,\cdots\cdots \tag{14}$$

The nonlinear filter equations developed in [67] were applied to this problem using simulated data having Gaussian random errors with standard deviations $\sigma_f = 100°F$ in the furnace flame temperature, $\sigma_w = 25°F$ in the furnace wall temperature, and $\sigma_s = 50°F$ in the ingot surface temperature measurements. The results are shown in Figure 2. The solid lines are the actual process while the dashed lines are the filter estimates. As can be seen, the filter performed very well even in the face of large process and measurement noise.

Similar problems of state estimation arise in slab reheating furnaces and other metallurgical operations. From the success of this simulation study and the experimental study [78] to be discussed later, the potential for future applications of these ideas to metals heating problems seems good.

2. Chemical Reactors

Chemical reactors freqently appear in tubular or other spatially distributed configurations, and often have unmeasurable state variables such as temperatures, concentrations, and catalyst activities. The general aspects of this problem have been discussed in great detail in [89], and various case studies may be found in [15,18-20,43-44,80,82,83,85]. As particular example, let us consider the problem of estimating the catalyst activity profile in a packed tubular chemical reactor having catalyst deactivation [82,83,85,89] (cf. Fig. 3). The problem arises because one can usually measure only the temperature and possibly the concentration profile in a packed tubular reactor, but would also like to know the state of the catalyst for reasons of feedback control, optimal planning of shutdowns, optimizing catalyst yields, etc.

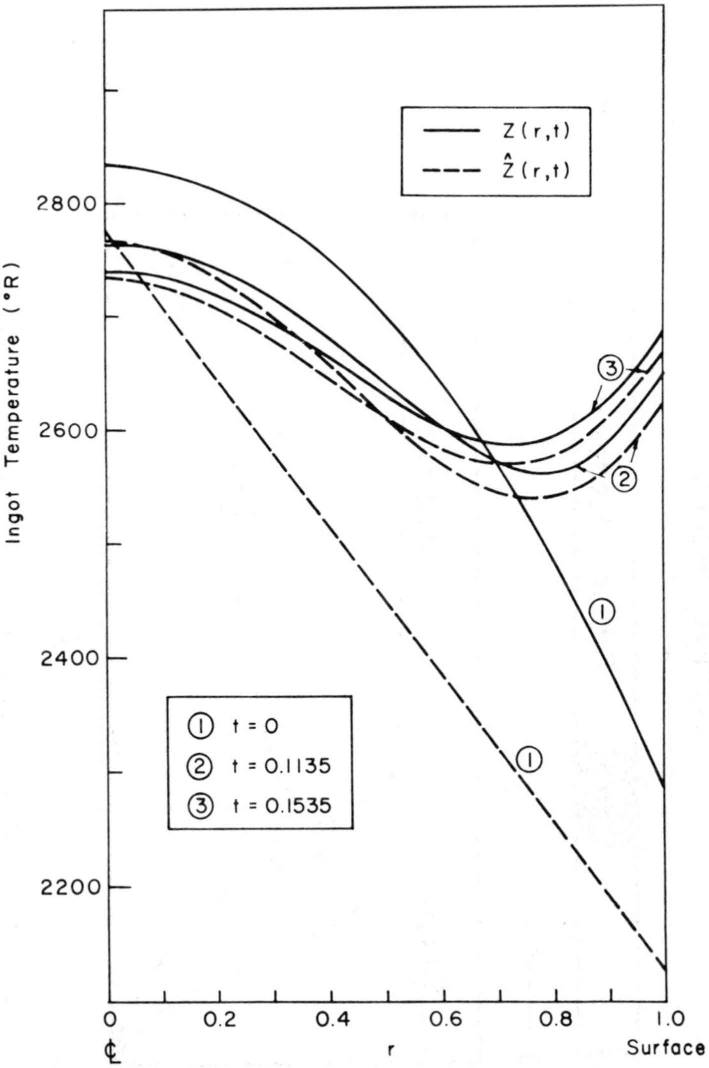

Figure 2. Performance of the nonlinear distributed parameter filter $\sigma_f = 100°F$, $\sigma_W = 25°F$, $\sigma_s = 50°F$.

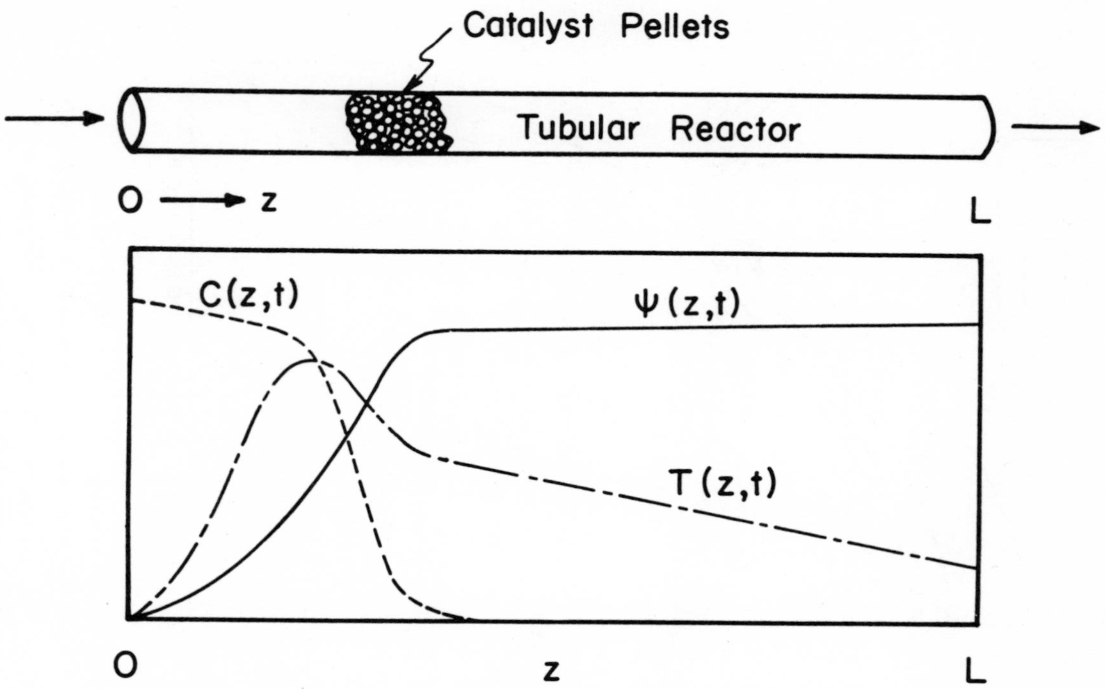

Figure 3. A packed bed tubular reactor with decaying catalyst

This problem was treated by simulation in [20,82,83], and an experimental case study, with off-line filter claculations, is reported in [85]. In this first experimental study of chemical reactor estimation, it was found that the filter, based on gas concentration measurements, improved the estimates of catalyst activity early in the catalyst lifetime and were consistent with the single direct activity measurement available at shutdown.

3. Air and Water Pollution

The monitoring of pollutant concentrations in urban air sheds or rivers and estuaries is a problem of current importance. Monitoring stations are costly and thus limited in number. Governmental agencies charged with maintaining air and water quality would therefore like to locate these stations to yield the maximum amount of information. In addition the state estimators must produce sufficiently reliable estimates in real time to allow effective enforcement of effluent quality laws and timely control action during episodes of substandard air and water quality.

In spite of the natural applicability of distributed parameter state estimation to these problems, the actual implementation has been very slow - principally due to the lack of adequate models. For air pollution, the modelling problem is particularly difficult because of the great number and unpredictability of model parameters and the complexity of the stochastic turbulence equations (cf. [90,91]). Nevertheless, a few papers have appeared applying state estimation to both air [86,92] and water pollution [75,76] problems. Hopefully with improved models, one will see the successful application of these ideas to a test region followed by the adoption of the use of state estimation on a routine basis by monitoring agencies.

4. Problems with Moving Boundaries

A class of systems for which there are many practical problems of current interest are those systems having moving boundaries. The theory has just been developed and some potential applications proposed [68]. These range from the estimation of temperature profiles and boundary position in melting and solidification problems found in process metallurgy to the estimation of the extent of crude oil spills.

A problem receiving current attention by the author is the estimation of the solid steel-mushy zone interface in the continuous casting of steel. The process, sketched in Figure 4,involves pouring molten steel at the top of a water cooled mold and continuously drawing out a thin-walled steel strand at the bottom. If the solid steel crust is too thin when leaving the mold, either due to some process upset or because the withdrawal rates are too high, the molten steel core will "break-out" and the castin machine must be shut down. By employing a distributed parameter filter to estimate the steel thickness in real time, one could operate at high average withdrawal rates while detecting potential break-outs before they occur and thus take appropriate control action. The performance of such a filter is being tested both numerically and experimentally at present.

There are a number of other areas of potential applications which shall not be discussed here. Some examples of these would be the estimation of the state of deformation of elastic structures [74] and the deciphering of signals in sonar detection [93].

III. A Real Time Case Study:

The first reported application of on-line distributed parameter state estimation and stochastic feedback control [78] involved estimating and subsequently controlling the temperature profile in a heated slab. The experimental apparatus is sketched in Figure 5. The slab is one meter long, 2 cm thick, and 25 cm wide. Heating of the slab is provided by 20 heating lamps oriented transversely both above and below the slab. Cooling is provided by water flowing through 20 holes through the center of the slab. There are 21 thermocouples located at the center of the slab at 21 positions along the length. The thermocouple signals are multiplexed and sent to the computer while heating control signals are sent from the computer. Further details of the equipment may be found in [78,94].

Because of the design of the apparatus, there are only significant temperature variations in the axial z direction. Thus one may model the system by

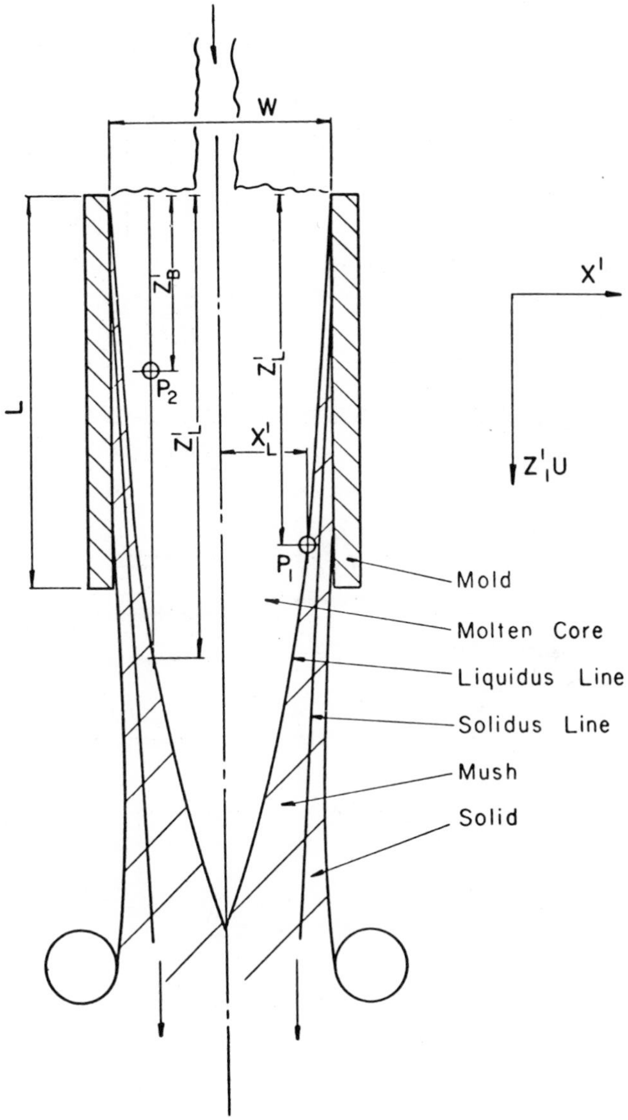

Figure 4. A schematic diagram of a continuous casting machine

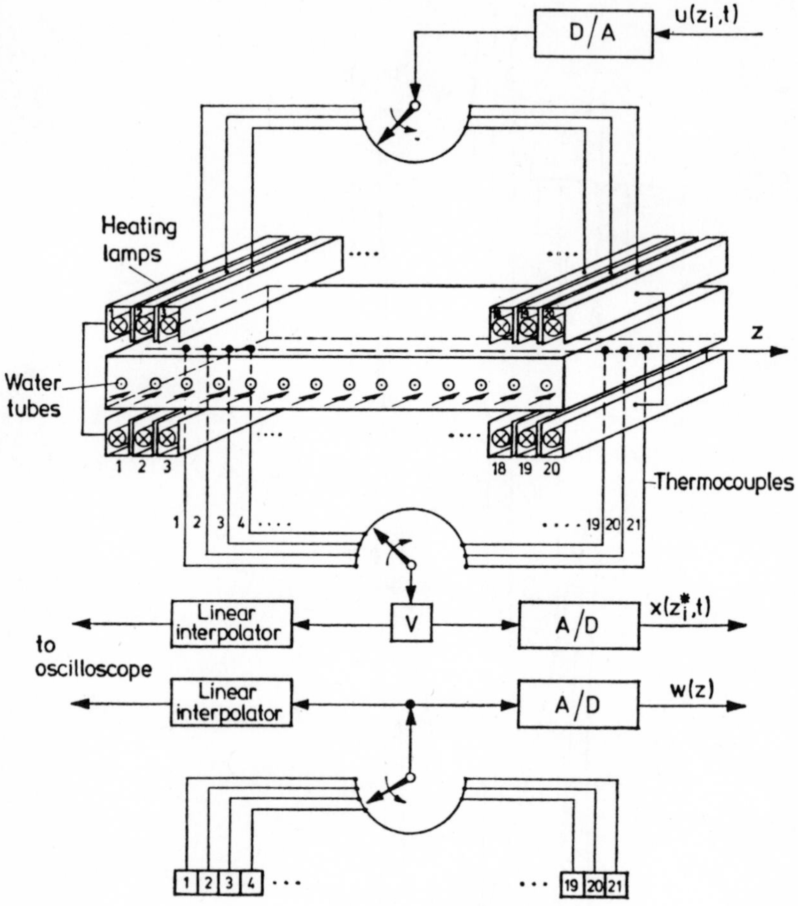

Figure 5. The heated slab used for state estimation and control studies

$$\frac{\partial x}{\partial t}(z,t) = \alpha \frac{\partial^2 x(z,t)}{\partial z^2} - \beta x(z,t) + \gamma u(z',t) \tag{15}$$

$$0 \le z \le 1$$

$$t \ge 0$$

$$z = 0,\ z = 1,\ \frac{\partial x}{\partial z} = 0 \tag{16}$$

where $x = T - T_w$, $\alpha = \dfrac{k}{\rho C_p \ell^2}$, $\beta = \alpha_w / \rho C_p$, $\gamma = C/\rho C_p$ (17)

The measuring device consisted of thermocouple signals taken at one or more points along the axis of the slab.

$$y_i(t) = x(z_i^*,t) + \eta_i(t) \qquad i = 1,2,\cdots m \tag{18}$$

where $\eta_i(t)$ represents the measurement error.

Because the system is linear, the filter equations take the standard form

$$\frac{\partial \hat{x}(z,t)}{\partial t} = \alpha \frac{\partial \hat{x}(z,t)}{\partial z^2} - \beta \hat{x}(z,t) + \gamma u(z,t)$$
$$+ \sum_{i=1}^{m} \sum_{j=1}^{m} P(z,z_i^*,t) Q_{ij}(t) [y_j(t) - \hat{x}(z_j^*,t)] \tag{19}$$

with boundary conditions (16) on the estimates. The covariance of the estimates may be computed from

$$\frac{\partial P(z,z',t)}{\partial t} = -2\beta P(z,z',t) + \alpha \left[\frac{\partial^2 P(z,z',t)}{\partial z^2} + \frac{\partial^2 P(z,z',t)}{\partial z'^2} \right]$$
$$- \sum_{i=1}^{m} \sum_{j=1}^{m} P(z,z_i^*,t) Q_{ij}(t) P(z_j^*,z',t) + R^+(z,z',t) \tag{20}$$

$$\frac{\partial P(z,z',t)}{\partial z} + \alpha\, R_0^{-1}(t)\delta(z') = 0 \qquad z = 0 \tag{21}$$

$$\frac{\partial P(z',z,t)}{\partial z'} + \alpha\, R_0^{-1}(t)\delta(z) = 0 \qquad z' = 0 \tag{22}$$

$$\frac{\partial P(z,z',t)}{\partial z} - \alpha\, R_1^{-1}(t)\delta(z'-1) = 0 \qquad z = 1 \tag{23}$$

$$\frac{\partial P(z',z,t)}{\partial z'} - \alpha\, R_1^{-1}(t)\delta(z-1) = 0 \qquad z' = 1 \tag{24}$$

The fact that the equations are linear makes the implementation through eigenfunction decomposition straightforward. By substitution of

$$\hat{x}(z,t) = \sum_{i=0}^{N} a_i(t)\varphi_i(z) \tag{25}$$

$$P(z,z',t) = \sum_{i=0}^{N} \sum_{j=0}^{N} B_{ij}(t)\varphi_i(z)\varphi_j(z') \tag{26}$$

$$u(z,t) = \sum_{i=0}^{N} u_i^*(t)\varphi_i(z) \tag{27}$$

into equations (19-24) yields a set of $(N+1)^2 + N+1$ ODE's in the Fourier coefficients $a_i(t)$, $B_{ij}(t)$:

$$\frac{d\underset{\sim}{a}(t)}{dt} = \underset{\sim}{\Lambda}a + \underset{\sim}{B}(t)\underset{\sim}{\theta}^T\underset{\sim}{Q}(t)[\underset{\sim}{y}-\underset{\sim}{\theta}\underset{\sim}{a}] + \underset{\sim}{u}^*(t) \tag{28}$$

$$\begin{aligned}\frac{d\underset{\sim}{B}(t)}{dt} = {} & \underset{\sim}{\Lambda}B(t) + \underset{\sim}{B}(t)\underset{\sim}{\Lambda} - \underset{\sim}{B}(t)\underset{\sim}{\theta}^T\underset{\sim}{Q}(t)\underset{\sim}{\theta}\underset{\sim}{B}(t) \\ & + \underset{\sim}{D}(t) + \underset{\sim}{D}_0(t) + \underset{\sim}{D}_1(t)\end{aligned} \tag{29}$$

The matrix Λ is a diagonal matrix of the first N eigenvalues corresponding to the first N eigenfunctions:

$$\varphi_0 = 1, \quad \varphi_i(z) = \cos i\pi z \qquad i = 1,2,\cdots N \tag{30}$$

The matrix $\underset{\sim}{\theta}$ is defined by

$$\underset{\sim}{\theta}^T = [\underset{\sim}{\varphi}(z_1*), \underset{\sim}{\varphi}(z_2*) \cdots \underset{\sim}{\varphi}(z_m*)] \tag{31}$$

while

$$\underset{\sim}{D}_0(t) = 2\alpha^2 \underset{\sim}{\varphi}(0)\underset{\sim}{R}_0^{-1}(t)\underset{\sim}{\varphi}^T(0) \tag{32}$$

$$\underset{\sim}{D}_1(t) = 2\alpha^2 \underset{\sim}{\varphi}(1)\underset{\sim}{R}_1^{-1}(t)\underset{\sim}{\varphi}^T(1) \tag{33}$$

and $D(t)$ is the matrix of coefficient arising from

$$R^+(z,z',t) = \sum_{i=0}^{N} \sum_{j=0}^{N} d_{ij}(t)\varphi_i(z)\varphi_j(z') \tag{34}$$

Due to the symmetry of the covariance equations only $\frac{(N+1)^2 + N+1}{2}$ ODE's for the $B_{ij}(t)$ need be solved (and these can be done off-line) along with N+1 ODE's for the estimates, $a_i(t)$. The computational effort may be summarized in tabular form as

Table 3

Computational effort for filter implementation

Number of Eigenfunctions needed, N+1	Estimate Eq'ns (on-line) (N+1)ODE's	Covariance Eq'ns (off-line) $\frac{(N+1)^2 + N+1}{2}$ ODE's	Total Eq'ns to be solved ODE's
2	2	3	5
3	3	6	9
4	4	10	14
5	5	15	20
6	6	21	27

In the present experimental study, it was found that $N+1 = 3$ was adequate to represent all the temperature profiles studied; thus a total of 9 ODE's had to be solved. For the sake of convenience all 9 ODE's were solved on-line in real time and yet the computational effort required was found to be less than 0.5% of real time.

The experimental results for this case study were very impressive (cf [78] for more complete details). Figure 6 shows typical open loop results when only a single temperature measurement at $z = 0$ was taken and a substantial random error added to the measurement before passing it to the filter. In this case a random error was taken from a Gaussian distribution having a standard deviation of 8°C. As can be seen, the initial condition given the filter was some 30°C in error, and yet the filter converges very quickly to within 1-2°C of the exact profile.

Typical closed loop results are shown in Figure 7. In this case the filter, having only one temperature measurement (at $z=0$) with 8°C std. deviation added measurement error, is coupled with a proportional + integral modal feedback controller making use of the certainty-equivalence principle. In all cases the stochastic feedback controller worked well with the estimates and exact profiles converging quickly and both reaching the desired set point. The results in Figure 7 are for an untuned filter and controller; thus one would expect that with tuning the rate of convergence of the stochastic controller could be improved still further.

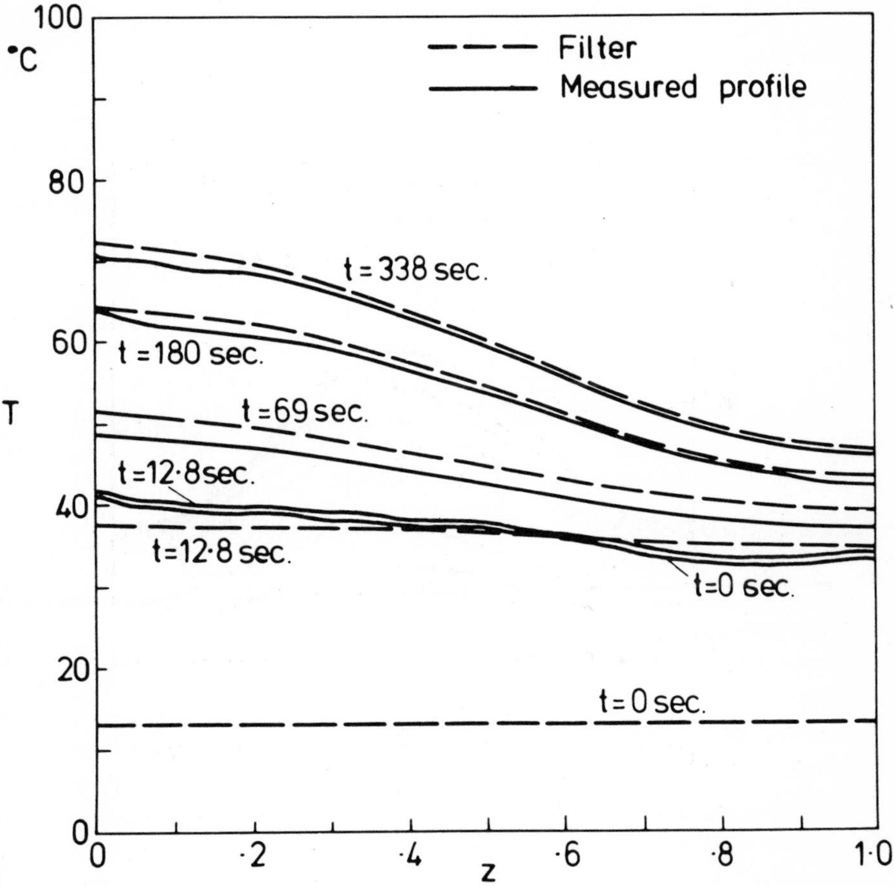

Figure 6. Open loop distributed parameter filter performance for the heated slab with one temperature sensor (z=0) and 8°C standard deviation added measurement errors.

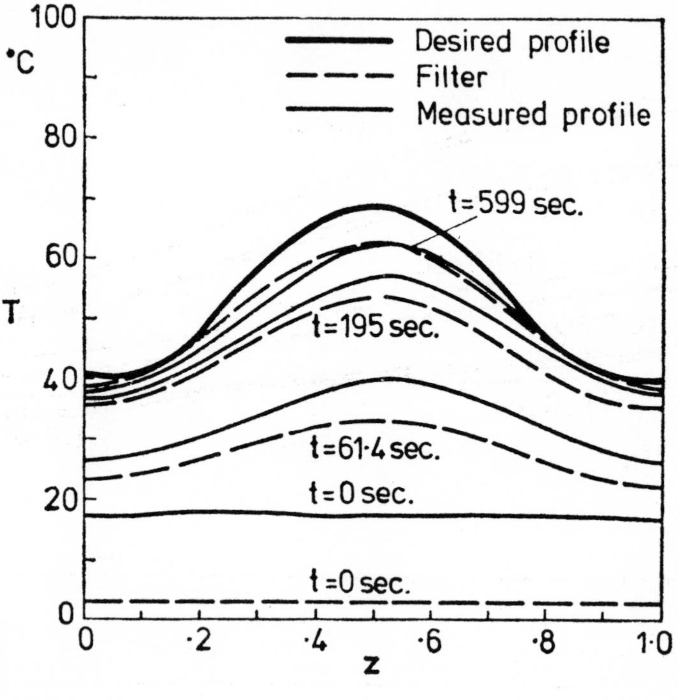

Figure 7. Closed loop distributed parameter filter and stochastic feedback controller performance for the heated slab having only one temperature sensor (z=0) and 8°C standard deviation added measurement errors.

IV. Concluding Remarks

From the discussion above it should be clear that there are a wide range of rigorous theoretical results available for optimal state estimation in linear distributed parameter systems. At the same time, a substantial number of approximate optimal filtering results leading to useful algorithms systems have appeared. Nevertheless, there are several areas in which further theoretical work would be of great help in applications. Some examples which spring to mind would be conditions for observability which extend to a wider range of systems and which are easy to apply. This would greatly aid in the choice of location for sensors. A second area of need is the development of an interaction/decomposition theory for stochastic feedback control which extends to a wider range of linear and nonlinear systems and to a wider range of error distributions. If such a theory were able to contain sensitivity information with regard to sensor and controller placement, this would be extremely helpful in choosing the optimal sensor location and controller application points.

In the applications/implementation sphere, further research is, perhaps, even more urgently needed. Our experimental study described in the last section (which seems to be the only experimental work yet to appear) has done much to show that distributed parameter state estimation and feedback control algorithms can be implemented on minicomputers if one is careful in choosing the numerical algorithms. Thus one pressing need is for work developing and demonstrating efficient algorithms for real time implementation. Our own research is proceeding in this direction. We currently have under development algorithms suitable for linear multi-dimensional systems as well as others for nonlinear systems in single or multiple space dimensions. All of these techniques are being applied to experimental model processes through the use of our PDP 11/40 minicomputer system and the results shall be reported elsewhere in due course.

As noted above, nearly all of the applications of distributed parameter state estimation and feedback control reported remain potential applications because of the lact of convincing demonstrations that the complex equations arising in real life problems can be feasibly solved on-line with minicomputers. It is our hope that the encouraging experience from these first experimental results will inspire the potential industrial and governmental user to begin applying these techniques to the solution of relevant real world problems. This in turn should provide feedback to the theoretician in the form of new challenging problems and thus bring enhanced vitality to the entire field.

REFERENCES

1. Rajbman, N. S., "The Application of Identification Methods in the USSR," Automatica, 12, 73 (1976).
2. Goodson, R. E. and M. Polis, "Identification of Parameters in Distributed Systems," Chapter in Identification, Estimation, and Control of Districuted Parameter Systems, edited by W. H. Ray and D. G. Lainiotis, Marcel Dekker (1976).
3. Kalman, R. E., "A New Approach to Linear Filtering and Prediction Problems," Trans. ASME, J. Basic Eng., 82, 35 (1960).
4. Kalman, R. E. and R. S. Bucy, "New Results in Linear Filtering and Prediction Theory," Trans. ASME, J. Basic. Eng., 83, 95 (1961)
5. Kwakernaak, H., "Optimal Filtering in Linear Systems with Time Delays," IEEE Trans. Auto. Cont. AC-12, 169 (1967).
6. Falb, P. H., "Infinite Dimensional Filtering: The Kalman-Bucy Filter in Hilbert Space," Information and Control, 11, 102 (1967).
7. Tzafestas, S. G. and J. M. Nightingale, "Optimal Filtering, Smoothing, and Prediction in Linear Distributed Parameter Systems," Proc. IEE, 115, 1207 (1968).
8. Tzafestas, S. G. and J. M. Nightingale, "Optimal Control of a Class of Linear Stochastic Distributed Parameter Systems," Proc. IEE, 115, 1213 (1968).
9. Tzafestas, S. G. and J. M. Nightingale, "Concerning Optimal Filtering Theory of Linear Distributed Parameter Systems," Proc. IEE, 115, 1737 (1968).
10. Tzafestas, S. G. and J. M. Nightingale, "Maximum Likelihood Approach to the Optimal Filtering of Distributed Parameter Systems," Proc. IEE, 116, 1085 (1969).
11. Balakrishnan, A. V. and J. L. Lions, "State Estimation for Infinite Dimensional Systems," J. Computer and System Sci., 1, 391 (1967).
12. Thau, F. E., "On Optimum Filtering for a Class of Linear Distributed Parameter Systems," Proceedings 1968 Joint Automatic Control Conference, p. 610; also appeared in Trans. ASME, J. Basic Eng., 91, 173 (1969)
13. Seinfeld, J. H., "Non-Linear Estimation for Partial Differential Equations," Chemical Engineering Science, 24, 75 (1969).
14. Ray, W. H., "Distributed Parameter State Estimation Algorithms and Applications - A Survey," Proceedings 6th IFAC Congress, paper 8.1 (1975).

15. Seinfeld, J. H., G. R. Gavalas and M. Hwang, "Non-Linear Filtering in Distributed Parameter Systems," Trans. ASME, J. Dyn. Sys. Meas. Cont., G93, 157 (1971).
16. Kuhr, D., "Optimale Filter für Lineare Systems Mit Verteilten Parametern," Regelungstechnik und Prozeβ-Datenverarbeitung, 18, 506 (1972).
17. Pell, T. M. and R. Aris, "Control of Linearized Distributed Systems on Discrete and Corrupted Observations," I and EC Fund, 9, 15 (1970).
18. Balchen, J. G., M. Field and T. O. Olson, "Multivariable Control with Approximate State Estimation of a Chemical Tubular Reactor," Proceedings IFAC Symposium on Multivariable Systems, Dusseldorf (1971), Paper 5.1.
19. Vakil, H. B., M. L. Michelson and A. S. Foss, "Fixed-Bed Reactor Control with State Estimation," I and EC Fund, 12, 328 (1973).
20. McGreavy, C. and A. Vago, "Estimation of Spatially Distributed Decay Parameters in a Chemical Reactor," Proceedings of IFAC Symposium on Identification and Syst. Parameter Est., Hague, 1973, p. 307.
21. Tzafestas, S. G., "Distributed Parameter State Estimation," Chapter in Identification, Estimation, and Control of Distributed Parameter Systems, edited by W. H. Ray and D. G. Lainiotis, Marcel Dekker (1976).
22. Wang, P. K. C., "Control of Distributed Parameter Systems," in Advances in Control Systems, Vol. 1, C. T. Leondes, Ed., Academic Press, 1964, p. 75.
23. Goodson, R. E. and R. E. Klein, "A Definition and Some Results for Distributed System Observability," IEEE Trans. Auto. Cont., AC-15, 165 (1970); also appeared in the Proceedings 1969 Joint Automatic Control Conference, p. 710.
24. Yu, T. K. and J. H. Seinfeld, "Observability of a Class of Hyperbolic Distributed Parameter Systems," IEEE Trans. Auto. Control, AC-16, 495 (1971).
25. Thowsen, A. and W. R. Perkins, "Observability Conditions for Two General Classes of Linear Flow Processes," IEEE Trans. Auto. Cont., AC-19, 603 (1974).
26. Sendaula, M., "Observability of Linear Systems with Time-Variable Delays," IEEE Trans. Auto. Cont., AC-19, 604 (1974).
27. Cannon, J. R. and R. E. Klein, "Optimal Selection of Measurement Locations in a Conductor for Approximate Determination of Temperature Distributions," Proceedings 1970 Joint Automatic Control Conference, p. 750.
28. Yu, T. K. and J. H. Seinfeld, "Observability and Optimal Measurement Location in Linear Distributed Parameter Systems," Int. J. Control, 18, 785 (1973).
29. Sakawa, Y. "Observability and Related Problems for Partial Differential Equations of Parabolic Type," SIAM J. Control, 13, 14 (1975).
30. Triggiani, R., "Extensions of Rank Conditions for Controllability and Observability to Banach Spaces and Unbounded Operators," SIAM J. Cont. and Opt., 14, 313 (1976).

31. Chen, W. H. and J. H. Seinfeld, "Optimal Location Process Measurements," Int. J. Control, 21, 1003 (1975).
32. Caravoni, P., G. DiPillo and L. DiPillo, "Optimal Location of a Measurement Point in a Diffusion Process," Proceedings 6th IFAC Congress, paper 8.3, (1975).
33. Aidarous, S. E., M. R. Gevers and M. J. Installe', "Optimal Sensor's Allocation Strategies for a Class of Stochastic Distributed Systems," Int. J. Control, 22, 197 (1975).
34. Koivo, A. J., "Optimal Estimator for Linear Stochastic Systems Described by Functional Differential Equations," Information and Control, 19, 232 (1971).
35. Koivo, A. J., "Optimal Control of Linear Stochastic Systems Described by Functional Differential Equations," J. of Opt. Theor. Appl., 9, 161 (1972).
36. Shukla, V. and M. D. Srinath, "Optimal Filtering in Linear Distributed Parameter Systems with Multiple Time Delays," Int. J. Control, 16, 673 (1972).
37. Padmanabhan, L., "On Filtering in Delay Systems with Continuous Dynamics and Discrete Time Observations," Proceedings 1972 Joint Automatic Control Conference, p. 826; also in AIChE J., 19, 517 (1973).
38. Lindquist, A., "A Theorem on Duality Between Estimation and Control for Linear Stochastic Systems with Time Delay," J. Math. Anal. Appl. 37, 516 (1972).
39. Koivo, H. N., "Least Squares Estimator for Hereditary Systems with Time Varying Delay," IEEE Trans. on Sys. Man. Cyb., SMC-4, 276 (1974).
40. Desai, P. R., K. S. Reddy and V. S. Rajamani, "Estimation in continuous Linear Time Delay Systems with Colored Observation Noise," Int. J. Sys. Sci., 6, 601 (1975).
41. Reddy, K. S. and V. S. Rajamani, "Optimal Estimation in Linear Distributed Systems with Time Delays," Int. J. Systems Sci., 6, 859 (1975).
42. Koivo, H. N and A. J. Koivo, "Control and Estimation of Systems with Time Delays," Chapter in Identification, Estimation, and Control of Distributed Parameter Systems, edited by W. H. Ray and D. G. Lainiotis, Marcel Dekker (1976).
43. Yu, T. K., J. H. Seinfeld and W. H. Ray, "Filtering in Non-Linear Time Delay Systems," IEEE Trans. Auto. Control, AC-19, 324 (1974).
44. Hwang, M., J. H. Seinfeld and G. R. Gavalas, "Optimal Least Square Filtering and Interpolation in Distributed Parameter Systems," J. Math. Anal. Appl., 39, 49 (1972).
45. Tzafestas, S. G., "Boundary and Volume Filtering of Linear Distributed Parameter Systems," Electronics Letters, 5, 199 (1969).
46. Phillipson, G. A. and S. K. Mitter, "State Identification of a Class of Linear Distributed Systems," Proc. Fourth IFAC Congress, Warsaw, Poland, June 1969.
47. Meditch, J. S., "On State Estimation for Distributed Parameter Systems," J. of The Franklin Institute, 290, 49 (1970).
48. Meditch, J. S., "Least Squares Filtering and Smoothing for Linear Distributed Parameter Systems," Automatica, 7, 315 (1971).

49. Kushner, H. J., "Filtering for Linear Distributed Parameter Systems," SIAM J. Cont., 8, 346 (1970).
50. Bensoussan, A., "Filtrage Optimal des Systèmes Linéaires," Dunod (1971).
51. Sakawa, Y1, "Optimal Filtering in Linear Distributed Parameter Systems," Int. J. Control, 16, 115 (1972).
52. Sholar, M. S. and D. M. Wiberg, "Canonical Equations for Boundary Feedback Control of Stochastic Distributed Parameter Systems," Automatica, 8, 287 (1972).
53. Kumar, G. R. V. and A. P. Sage, "Error Analysis Algorithms for Distributed Parameter Filtering," Automatica, 8, 363 (1972).
54. Atre, S. R. and S. S. Lamba, "Derivation of an Optimal Estimator for Distributed Parameter Systems Via Maximum Principle," IEEE Trans. Auto. Cont., AC-17, 388 (1972).
55. Atre, S. R. and S. S. Lamba, "Optimal Estimation in Distributed Processes Using the Innovations Approach," IEEE Trans. Auto. Cont., AC-17, 710 (1972).
56. Atre, S. R., "A Note on the Kalman-Bucy Filtering for Linear Distributed Parameter Systems," IEEE Trans. Auto. Cont., AC-17, 712 (1972).
57. Bhagavan, B. K. and L. R. Nardizzi, "Suboptimal Stochastic Control of a Class of Linear Distributed Parameter Regulators," Proceedings 1972 Joint Automatic Control Conf., p. 299.
58. Tzafestas, S. G., Bayesian Approach to Distributed-Paramter Filtering and Smoothing," Int. J. Control, 15, 273 (1972).
59. Tzafestas, S. G., "On Optimum Distributed Parameter Filtering and Fixed-Interval Smoothing for Colored Noise," IEEE Trans. Auto. Cont., AC-17, 448 (1972).
60. Sherry, H. and D. W. C. Shen, "Combined State and Parameter Estimation for Distributed Parameter Systems Using Discrete Observatons," Proceedings IFAC Symposium on Identification and System Parameter Estimation, Hague (1973) p. 71.
61. Phillipson, G. A., "Identification of Distributed Systems," American Elsevier, New York (1971).
62. Tzafestas, S. G., "On the Distributed Parameter Least Squares State Estimation Theory," Int. J. Sys. Sci., 4, 833 (1973).
63. Curtain, R. L.,"Infinite Dimensional Filtering", SIAM J. Control, 13, 89, (1975).
64. Reddy, K. S. and V. S. Rajamani, "Optimal Filtering for a Class of Distributed Parameter Systems with Colored Measurement Noise," Int. J. Systems Sci., 6, 615 (1975).
65. Seinfeld, J. H. and M. Hwang, "Some Remarks on Non-Linear Filtering in Distributed Parameter Systems," Chem. Eng. Sci., 25, 741 (1970).
66. Lamont, G. B. and K. S. P. Kumar, "State Estimation in Distributed Parameter Systems Via Least Squares and Invariant Imbedding," J. Math. Anal. Appl., 38, 588 (1972).
67. Ajinkya, M. B., W. H. Ray, T. K. Yu and J. H. Seinfeld, "The Application of an Approximate Non-Linear Filter to Systems Governed by Coupled Ordinary and Partial Differential Equations," Int. J. System Sci., 6, 313 (1975).

68. Ray, W. H. and J. H. Seinfeld, "Filtering in Distributed Parameter Systems with Moving Boundaries," Automatica, 11, 509 (1975).
69. Bensoussan, A., "On the Separation Principle for Distributed Parameter Systems," Proceedings IFAC Symposium on Distributed Paramter Systems, Banff (1971), paper 8.1.
70. Bensoussan, A., "Control of Stochastic Partial Differential Equations," chapter in Identification, Estimation, and Control of Distributed Parameter Systems, edited by W. H. Ray and D. G. Lainiotis, Marcel Dekker (1976).
71. Hewer, G. A. and G. J. Nazaroff, "Observer Theory for Delayed Differential Equations," Int. J. Control, 18, 1 (1973).
72. Gressang, R. V. and G. B. Lamont, "Observers for Systems Characterized by Semigroups" IEEE Trans. Auto. Control, AC-20, 523 (1975).
73. Bhat, K. P. M. and H. N. Koivo, "An Observer Theory for Time Delay Systems," Control Systems Report 7502, University of Toronto (1975).
74. Köhne, M., "The Control of Vibrating Elastic Systems," chapter in Identification, Estimation, and Control of Distributed Parameter Systems, edited by W. H. Ray and D. G. Lainiotis, Marcel Dekker (1976).
75. Perlis, H. J. and B. Okunseinde, "Multiple Kalman Filters in a Distributed Stream Monitoring System," Proceedings 1974 Joint Automatic Control Conference, p. 615.
76. Özgören, M. K., R. W. Longman and C. A. Cooper, "Stochastic Optimal Control of Artificial River Aeration," Proceedings 1974 Joint Automatic Control Conference, p. 235.
77. Hassan, M. A., M. A. R. Ghonaimy and M. A. Abd El-shaheed, "A Computer Algorithm for Optimal Discrete Time State Estimation of Linear Distributed Parameter Systems," Proceedings IFAC Symposium on Dist. Para. Syst., Banff (1971), paper 13.7.
78. Ajinkya, M. B., M. Köhne, H. F. Mäder and W. H. Ray, "The Experimental Implementation of a Distributed Parameter Filter," Automatica, 11, 571 (1975).
79. Johnson, R. A., "Functional Equations, Approximations, and Dynamic Response of Systems with Variable Time Delay," IEEE Trans. Auto. Cont., AC-17, 398 (1972).
80. Yu, T. K. and J. H. Seinfeld, "Control of Stochastic Distributed Parameter Systems," Journ. Opt. Theor. Appl., 10, 362 (1972).
81. Yu, T. K. and J. H. Seinfeld, "Suboptimal Control of Stochastic Distributed Parameter Systems," AIChE J., 19, 389 (1973).
82. Goldman, S. F. and R. W. H. Sargent, "Applications of Linear Estimation Theory to Chemical Processes: A Feasibility Study," Chem. Eng. Sci., 26, 1535 (1971).
83. Joffe, B. L. and R. W. H. Sargent, "The Design of an On-Line Control Scheme for a Tubular Catalytic Reactor," Trans. Inst. Chem. Engr., 50, 270 (1972).
84. Maslowski, A., "Optimal Estimation for Space-Time Reactor Processes," Nuclear Science and Eng., 52, 274 (1973).
85. Ajinkya, M. B., W. H. Ray and G. F. Froment, "On-Line Estimation of Catalyst Activity Profiles in Packed Bed Reactors Having Catalyst Decay," I and EC Process Design and Development, 13, 107 (1974).

86. Bankoff, S. G. and E. L. Hanzevack, "The Adaptive Filtering Transport Model for Prediction and Control of Pollutant Concentration in an Urban Airshed," Atmospheric Environment, 9, 793 (1975).
87. Viskanta, R., R. E. Chupp, P. J. Hommert and J. S. Toor, "Thermal Remote Sensing of Temperature Distribution in Glass," Automatica, 11, 409 (1975).
88. Stark, K., "Regelung der ortsabhangigen Neutronendichte in einem Kernreaktor," Regelungstechnik, 23, 217 (1975).
89. Ray, W. H., "An Optimal Control Scheme for Tubular Chemical Reactors Suffering Catalyst Decay," Proceedings IFAC Symposium on Distributed Parameter Systems, Banff (1971), paper 10.5.
90. Seinfeld, J. H., "Current Problems in the Modelling and Control of Urban Air Pollution," Proceedings 1972 Joint Automatic Control Conference, p. 717.
91. Seinfeld, J. H., "Determination of Air Pollution Control Strategies," Proceedings 1974 Joint Automatic Control Conference, p. 554.
92. Gould, L. A., F. C. Schweppe and A. A. Desalu, "Dynamic Estimation of Air Pollution," Proceedings 1974 Joint Automatic Control Conference, p. 3.
93. Van Trees, H. L., "Applications of State Variable Techniques in Detection Theory," Proceedings IEEE, 58, 653 (1970).
94. Mäder, H. F., "Zeitoptimale steuerung und modale regelung eines technisch realisierten wärmeleitsystems," Ph.D. Thesis, University of Stuttgart (1975).

"NUMERICAL SOLUTION OF THE TRANSONIC EQUATION BY THE FINITE ELEMENT METHOD VIA OPTIMAL CONTROL"

M. O. Bristeau, R. Glowinski, O. Pironneau

ABSTRACT

It is shown that the transonic equation for compressible potential flow is equivalent to an optimal control problem of a linear distributed parameter system. This problem can be discretized by the finite element method and solved by a conjugate gradient algorithm. Thus a new class of methods for solving the transonic equation is obtained. It is particularly well adapted to problems with complicated two or three dimensional geometries and shocks.

1. PLAN

2. Introduction
3. Statement of the problem
4. Gelder's algorithm for subsonic flow
5. Formulation via optimal control
6. Discretization and numerical solution
7. Numerical results
8. Conclusion

2. INTRODUCTION

The transonic equation is a non linear partial differential equation which has an elliptic behavior in the subsonic regions of the flow and a hyperbolic behavior in the supersonic regions. At the interface the normal component of the speed of the flow can be discontinuous (shocks). Some finite difference methods have been successfully developed even for flows around simple 3-D objects (Jameson (1974), Garabedian-Korn (1971)). However the method of finite differences is not well suited to complicated geometries. An alternative approach using finite elements was studied by Gelder (1971), Norries and de Vries (1973), Periaux (1975) but their methods explode at supersonic speeds. Following Gelder's approach we shall replace the transonic equation by the minimization of a functional in an abstract space, a problem which can be solved by the methods of the theory of calculus of variations and optimal control theory.

3. STATEMENT OF THE PROBLEM

Stationary adiabatic monophasic compressible flows, in which the effects of viscosity are neglected, are well described by the set of equations

$$(3.1) \qquad \nabla\cdot(\rho u) = 0 \quad \left(\frac{\partial \rho u_1}{\partial x_1} + \frac{\partial \rho u_2}{\partial x_2} + \frac{\partial \rho u_3}{\partial x_3} = 0\right)$$

$$(3.2) \qquad \rho = \rho_0 \left(1 - \frac{\gamma-1}{\gamma+1}\frac{|u|^2}{C_*^2}\right)^{\frac{1}{\gamma-1}}$$

$$(3.3) \qquad u = \nabla\phi \; (u_i = \frac{\partial\phi}{\partial x_i}\,, \quad i = 1,2,3)$$

where ρ is the density, u is the speed of the fluid and where ρ_0, C_* and γ are constants (γ=1.4 for di-atomic gas, see Landau-Lifschitz (1971). We shall denote $k = \frac{\gamma-1}{\gamma+1}\frac{1}{C_*^2}$, $\alpha = 1/(\gamma-1)$. Therefore, if Ω is the region occupied by the fluid, one must solve the nonlinear partial differential equation:

$$(3.4) \qquad \nabla\cdot(1-k|\nabla\phi|^2)^{\alpha}\nabla\phi = 0 \quad \text{in } \Omega$$

with the boundary conditions

$$(3.5) \qquad \phi|\Gamma_1 = \phi_1$$

$$(3.6) \qquad \frac{\partial\phi}{\partial n}|\Gamma_2 = g_2$$

where Γ_1 and Γ_2 are parts of the boundary $\partial\Omega$ of Ω. We shall assume that $\Gamma_1 \cup \Gamma_2 = \partial\Omega$ and $\Gamma_1 \cap \Gamma_2 = 0$. In addition, if there are shocks (i.e. lines or surfaces where the tangential speed of the flow is continuous but the speed normal to these lines or surfaces is discontinuous) then, across the shock:

(3.7) $(\rho u)^+ = (\rho u)^-$ (Rankine-Hugoniot condition)

(3.8) $u_n^+ \leq u_n^-$ (entropy condition)

where it is understood that the particles of the fluid move from - to +.

Note the (3.4) multiplied by $w \in C^1(\Omega)$ and integrated by parts, leads to

(3.9) $$\int_\Omega (1-k|\nabla\phi|^2)^\alpha \nabla\phi \; \nabla w dx = \int_{\Gamma_2} (1-k|\nabla\phi|^2)^\alpha g_2 w d\Gamma_2$$

$$\forall w \in C^1(\Omega) \text{ s.t. } w|\Gamma_1 = 0 \; ; \; \phi|\Gamma_1 = 0$$

If the notion of derivative is extended and the space $C^1(\Omega)$ is replaced by $H^1(\Omega) = \{w \in L^2(\Omega) \; | \nabla w \in (L^2(\Omega))^3\}$ then (3.9) is called a weak formulation of (3.4)-(3.6). Note that it contains (3.7).

4. GELDER'S ALGORITHM FOR SUBSONIC FLOW

For notational convenience we suppose $g_2|\Gamma_2 = 0$. Consider the functional

(4.1) $$E_0(\phi) = - \int_\Omega (1-k|\nabla\phi|^2)^{\alpha+1} dx$$

we shall say that ϕ is a stationary point of E_0 on

(4.2) $$H^1_{o1}(\Omega) = \{\phi \in H^1(\Omega) | \phi| \; \Gamma_1 = 0\}$$

if

$$\delta E_0 = E_0(\phi+\delta\phi) - E_0(\phi) = o(\delta\phi) \; \forall \delta\phi \in H^1_{o1}(\Omega)$$

Since, from (4.1)

$$\delta E_0 = \int_\Omega 2k(\alpha+1)(1-k|\nabla\phi|^2)^\alpha \nabla\phi\nabla\delta\phi dx + o(\delta\phi) \tag{4.3}$$

any stationary point of E_0 on $H^1_{01}(\Omega)$ satisfies

$$\int_\Omega (1-k|\nabla\phi|^2)^\alpha \nabla\phi\nabla w dx = 0 \quad \forall w \in H^1_{01}(\Omega)$$

Thus all stationary points of E_0 on $H^1_{01}(\Omega)$ such that $\phi|\Gamma_1 = \phi_0|\Gamma_1$ and which satisfy (3.8) are solutions of our problem.

Let us look at

$$\frac{d^2}{d\lambda^2}(E_o(\phi+\lambda\delta\phi))|_{\lambda=0} = 2k(\alpha+1)\int_\Omega (1-k|\nabla\phi|^2)^\alpha [\nabla\delta\phi\nabla\delta\phi - \frac{2k\alpha(\nabla\phi\cdot\nabla\delta\phi)^2}{(1-k|\nabla\phi|^2)}] dx$$

with our notation the mach number is such that

$$M^2 = 2k\alpha(1-k|\nabla\phi|^2)^{-1} |\nabla\phi|^2$$

therefore, if θ is the angle between $\nabla\phi$ and $\nabla\delta\phi$;

$$\frac{d^2E_0}{d\lambda^2} = -2k(\alpha+1)\int_\Omega \rho(1-M^2\cos^2\theta)|\nabla\delta\phi|^2 dx$$

This shows that if in some part of the fluid $M > 1$, E is not convex and the solution of (3.4)-(3.8) is only a saddle point of E. On the other hand, if $M < 1$ in Ω then E is convex and the solution of (3.4)-(3.8) is a minimum of E . This fact was utilized by Gelder (1971) and Periaux (1975) for constructing a solution of (3.4)-(3.8). The functional E is minimized by a gradient method with respect to the $H^1(\Omega)$-norm ; i.e. $\{\phi_n\}_{n\geq 2}$ is constructed by solving for $\phi_{n+1} \in H^1(\Omega)$:

$$\int_\Omega \rho_n \nabla\phi_{n+1} \nabla w d\Omega = 0 \quad \forall w \in H^1_{01}(\Omega) \ , \ (\phi_{n+1}-\phi_1)|\Gamma_1 = 0$$

This method works very well (less than 15 iterations in most cases) and it is desirable to construct a method as near to it as possible, for supersonic flows.

5. FORMULATION VIA OPTIMAL CONTROL

Along the line of §5 we shall look for functionals which have the solution of (3.4)-(3.8) for minimum. Several functionals were studied in Glowinski-Pironneau (1975) and Glowinski-Periaux-Pironneau (1976). In this presentation we shall study the following functional:

$$(5.1)\qquad E(\xi) = \int_\Omega \rho(|\nabla\xi|^2)|\nabla(\phi-\xi)|^2\,dx,\quad \rho(|\nabla\xi|^2) = (1-k|\nabla\xi|^2)^\alpha$$

where $\phi = \phi(\xi)$ is the solution in $H^1(\Omega)$ of

$$(5.2)\qquad \int_\Omega \rho(|\nabla\xi|^2)\,\nabla\phi\nabla w dx = 0 \quad \forall w \in H^1_{01}(\Omega)\ ,\ \phi|\Gamma_1 = \phi_1$$

Proposition 1. Suppose that (3.4)-(3.8) has a solution.

Given $\varepsilon > 0$, small, the problem

$$(5.3)\qquad \min\ \{E(\xi)\,|\,\xi \in \Xi\}$$

where $\Xi = \{\xi \in H^1(\Omega)\,|\ \xi|\Gamma_1 = \phi_1 \quad |\nabla\xi(x)| \le k^{-1/2}(1-\varepsilon) \quad \text{a.e } x \in \Omega\}$

has at least one solution and if $\Delta\xi(x) < +\infty \quad \forall x \in \Omega$, it is a solution of (3.4)-(3.8). Furthermore any minimizing sequence $\{\xi_n\}_{n\ge 0}$ of (5.3) has a subsequence which satisfies (3.5)-(3.7) and

$$\lim_{n\to+\infty} \int_\Omega^n (1-k|\nabla\xi_n|^2)^\alpha \nabla\xi_n\ \nabla w dx = 0 \quad \forall w \in H^1_{01}(\Omega)$$

Proof: the first part of the theorem is obvious.

Let $\{\xi_n\}$ be a minimizing sequence of E then $\xi_n \in \Xi$ implies that $\|\nabla\xi_n\|^2 < k^{-1}(1-\varepsilon)^2 \int_\Omega dx$, therefore a subsequence (denoted $\{\xi_n\}$ also) converging towards a $\bar\xi \in \Xi$ can be extracted. Furthermore $\|\nabla(\phi_n-\xi_n)\| \to 0$. Therefore

$$\int_\Omega \rho_n\ \nabla(\phi_n-\xi_n)\ \nabla w dx = \int_\Omega \rho_n\ \nabla\xi_n\ \nabla w dx \to 0$$

for every subsequence such that ρ_n converges in the $L^\infty(\Omega)$ weak star topology.

Remark. Note that if $\bar{\xi}$ is a weak limit of $\{\xi_n\}$, $\bar{\xi}$ may not be a solution of (5.3). This, however, does not seem to create problems in practice.

Proposition 2

If $\xi|\Gamma_1 = \phi_1 \quad \delta\xi|\Gamma_1 = 0$, then

(5.6)

$$E(\xi+\delta\xi)-E(\xi) = 2\int_\Omega \rho(|\nabla\xi|^2)(1+\tfrac{1}{2}M^2(1-|\nabla\xi|^2\cdot|\nabla\phi|^{-2}))\nabla\xi\cdot\nabla\delta\xi dx+o(\delta\xi)$$

$$(M^2 = -2\rho'\rho^{-1}|\nabla\phi|^2 = +2k\alpha(1-k|\nabla\phi|^2)^{-1}|\nabla\phi|^2)$$

Proof

From (5.1) and (5.2)

(5.7)

$$E(\xi+\delta\xi)-E(\xi) = 2\int_\Omega[2\rho'\nabla\xi\cdot\nabla\delta\xi|\nabla(\phi-\xi)|^2-\rho\nabla(\phi-\xi)\nabla\delta\xi+\rho\nabla(\phi-\xi)\cdot\nabla\delta\phi]dx$$

$$+ o(\delta\xi) + o(\delta\xi)$$

where

$$\rho' = -k\alpha(1-k|\nabla\xi|^2)^{\alpha-1}$$

From (5.3)

(5.8) $$\int_\Omega \rho\nabla\delta\phi\nabla w dx = -\int_\Omega 2\rho'\nabla\xi\cdot\nabla\delta\xi\nabla\phi\cdot\nabla w dx+o(\delta\xi) \quad \forall w \in H^1_{01}(\Omega)$$

and since $\rho(|\nabla(\xi+\delta\xi)|^2)$ is bounded from below by a positive number, there exists K such that $\|\nabla\delta\phi\| \le K\|\nabla\delta\xi\|$, if $\alpha > 2$.

Therefore, by letting $w = \phi-\xi$ in (5.8), (5.7) becomes

$$\delta E = -2\int_\Omega [\rho\nabla(\phi-\xi)\cdot\nabla\delta\xi + \rho'(|\nabla\phi|^2-|\nabla\xi|^2)\nabla\xi\cdot\nabla\delta\xi]\, dx$$

and from (5.2) the term $\rho\nabla\phi\nabla\delta\xi$ disappears.

Corollary 1

If $\bar{\xi}, \bar{\phi}$ is a stationary point of E, it satisfies:

$$\nabla \cdot [\bar{\rho}(1 + \frac{\bar{M}^2}{2}(1 - |\nabla \bar{\xi}|^2 |\nabla \bar{\phi}|^{-2}) \nabla \bar{\xi}] = 0 \quad \text{in} \quad \Omega \tag{5.9}$$

$$\bar{\rho}(1 + \frac{\bar{M}^2}{2}(1 - |\nabla \bar{\xi}|^2 |\nabla \bar{\phi}|^{-2}) \frac{\partial \bar{\xi}}{\partial n}|_{\Gamma_2} = 0 \; ; \; \bar{\xi}|\Gamma_1 = \phi_1 \tag{5.10}$$

Remark: It should be noted that in most cases (5.3) has no other stationary point than the solutions of (3.4)-(3.7). Indeed let (x_ξ, y_ξ, z_ξ) be a curvilinear system of coordinate such that

$$\nabla \xi = (\frac{\partial \xi}{\partial x_\xi}, 0, 0)$$

Then, from (5.9), (5.10)

$$\frac{\partial}{\partial x_\xi}[\bar{\rho}(1 + \frac{\bar{M}^2}{2}(1 - |\nabla \bar{\xi}|^2 |\nabla \bar{\phi}|^{-2}) \frac{\partial \xi}{\partial x_\xi}] = 0, \; \frac{\partial \bar{\xi}}{\partial n}|_{\Gamma 2} = 0 \tag{5.11}$$

$$\text{or} \quad \bar{M}^2(1 - |\nabla \bar{\xi}|^2 |\nabla \bar{\phi}|^{-2})|_{\Gamma_2} = -2, \; \bar{\xi}|\Gamma_1 = \phi_1$$

This system looks like the one dimensional transonic equation for a compressible fluid with density

$$\bar{\rho}(1 + \frac{\bar{M}^2}{2}(1 - |\nabla \bar{\xi}|^2 |\nabla \bar{\phi}|^{-2}))$$

Therefore, if the ξ-stream lines meet two boundaries and $\Delta \xi < +\infty$ at the shocks and

$$1 + \frac{\bar{M}^2}{2}(1 - |\nabla \bar{\xi}|^2 |\nabla \bar{\phi}|^{-2}) > 0$$

then $\bar{\phi} = \bar{\xi}$.

6. DISCRETIZATION AND NUMERICAL SOLUTIONS

Let τ_h be a set of triangles or tetrahedra of Ω where h is the length of the greatest side. Suppose that

$$\bigsqcup_{T\in\tau_h} T \subset \Omega \ , \ T_1 \cap T_2 = \quad \text{or a vertex} \quad \forall T_1, T_2 \in \tau_h \ ,$$

Let $\Omega_h = \overset{\circ}{\overline{\bigsqcup_{T\in\tau_h} T}}$ and Γ_{1h}, Γ_{2h} parts of $\partial\Omega_h$ which approximate Γ_1 and Γ_2.

Let H_h be an approximation of $H^1(\Omega)$:

(6.1) $$H_h = \{w_h \in C^0(\Omega_h) \,|\, w_h \text{ linear on } T \ \forall T \in \tau_h\}$$

Note that any element of H_h is completely determined by the values that it takes at the nodes of τ_h. Therefore if we assume that τ_h has $N = n+p+m$ nodes P_i with $P_i \in \Gamma_{1h}$ if $i > n+p$, $P_i \in \Gamma_{2h}$ if $i \in]n,n+p[$, and if we define $w_i \in H_h$ by

(6.2) $$w_i = 1 \text{ at node 1 and zero at all other nodes}$$

Then any function $w \in H_h$ is written as

(6.3) $$\phi = \Sigma\alpha_i w_i$$

Algorithm 1

Let $\xi_h = \sum_{i=1}^{N} \xi^i w_i$, then (5.2) becomes

(6.4)
$$\int_\Omega (1-k|\nabla\xi_h|^2)^\alpha \nabla\phi_h \nabla w_i \, dx = 0 \qquad i=1,\dots,n+p$$
$$\phi_h = \sum_{i=1}^{n+p} \phi^i w_i + \sum_{n+p+1}^{N} \phi_1^i w_i$$

and (5.6) becomes

$$\frac{1}{2}\,\delta E_h = \sum_{i=1}^{N} \delta\xi^i \delta E_h^i + o(\delta\xi^i) \tag{6.5}$$

$$\delta E_h^i = \int_\Omega [\rho - \rho'(|\nabla\phi_h|^2 - |\nabla\phi_h|^2)]\nabla\xi_h \cdot \nabla w_i \; dx \tag{6.6}$$

Consider the following algorithm

<u>Step 0</u> Choose τ_h, ξ_{ho} set $j=0$

<u>Step 1</u> Compute ϕ_{hj} by solving (6.4) with $\xi_h = \xi_{hj}$

<u>Step 2</u> Compute $\{\delta E_{hj}^i, \; i=1,\ldots,N\}$ by (6.6)

<u>Step 3</u> Compute $\delta\xi_h = \sum_{i=1}^{n+p} \delta\xi^i w_i$ by solving

$$\int_{\Omega_h} \nabla\delta\xi_h \nabla w_i dx = \delta E_{hj}^i, \; i=1,\ldots,n+p \tag{6.7}$$

<u>Step 4</u> Compute an approximation $\bar{\lambda}_j$ of the solution of

$$\min_{\lambda\in[0,1]} \int_{\Omega_h} \rho(\lambda)|\,\nabla(\xi_h(\lambda) - \phi_h(\lambda))|^2 dx \tag{6.8}$$

where

$$\xi_h(\lambda) = \sum_{i=1}^{N} (\xi_{hj}^i - \lambda\delta\xi_h^i) w_i$$

<u>Step 5</u> Set $\xi_{h_{j+1}} = \xi_h(\bar{\lambda}_j)$, $j=j+1$ and go to step 1.

<u>Proposition 3</u>

Let $\{\xi_{hj}\}_{j\geq 0}$ be a sequence generated by algorithm 1 such that $|\nabla\xi_{hj}(x)| \leq k^{-1/2}$ $\forall x$, $\forall j$. Every accumulation point of $\{\xi_{hj}\}_{j\geq 0}$ is a stationary point of the functional

$$E_h(\xi_h) = \int_{\Omega_h} |\nabla(\phi_h - \xi_h)|^2 dx \tag{6.9}$$

where $\phi_h = \phi_h(\xi_h)$ is the solution of (6.4), in

$$\Xi_h = \{\xi_h \in H_h \mid |\nabla\xi_h(x)| \leq k^{-1/2} \ \forall x \in \Omega_h\}$$

Proof

Algorithm 1 is the method of steepest descent applied to minimize (6.9) in Ξ_h , with the norm

(6.10) $$\|\xi_h\|_h^2 = \int_{\Omega_h} \nabla\xi_h \nabla\xi_h \, dx$$

Therefore $\{E_h(\xi_{hj})\}_j$ decreases until δE_{hj} reaches zero.

Remark 6.1: (6.4) should be solved by a method of relaxation but (6.7) can be factorized once and for all by the method of Choleski.

Remark 6.2: Problem (6.8) is usually solved by a Golden section search or a Newton method.

Remark 6.3: Step 5 can be modified so as to obtain a conjugate gradient method.

Remark 6.4: The restriction: $|u_{h_j}(x)| \leq k^{-1/2}$ in theorem 5.1 is not a problem if u is not too close to $k^{-1/2}$, otherwise one must treat this restriction as a constraint in the algorithm. Also, even though theorem (5.1) ensures the computation of stationary points only, it is a common experience that global minima can be obtained by this procedure if there is a finite number of local minima.

Remark 6.5: The entropy condition $\Delta\xi_h < +\infty$ can be taken into account numerically. Let $M(x)$ be a real valued function then $\Delta\xi_h \leq M(x)$ becomes, from (6.7)

(6.11) $$-\Sigma\, \bar{\lambda}_j \, \delta E^i_{hj} \leq M(x_i) \quad i = 1,\ldots,n+p$$

Therefore, to satisfy (6.11) at iteration $j+1$, it suffices to take $\delta E^i_{hj} = 0$ in (6.7) for all i such that (6.11) at iteration j is an equality. This procedure amounts to control $\omega = \Delta\xi$ instead of ξ .

7. NUMERICAL RESULTS

The method was tested on a nozzle discretized as shown on figure 1, (300 triangular elements, 180 nodes). The Polak-Ribiere method of conjugate gradient was used with an initial control: $\Delta\xi = 0$ (incompressible flow). A mono-dimensional optimization subroutine based on a dichotomic search was given to us by Lemarechal. Several boundary conditions were tested -

1°) Subsonic mach number $M_\infty = 0.63$ at the entrance, zero potential on exit, the method had already converged in 10 iterations (to be compared with the Gelder-Periaux method) giving a criterion $E_{h10} = 2\ 10^{-13}$ $(E_{ho} = 10^{-4})$.

2°) Entrance and exit potential specified.

For a decrease of potential of $\phi_1 - \phi_2 = 0.7$ the method had converged in 20 iterations without including the entropy condition, giving a criterion of $E_{h20} = 5\cdot 10^{-7}$, the results are shown on figure 2.

3°) Supersonic mach number $M_\infty = 1.25$

The method computes a solution that has a shock at the first section of discretization. Another boundary condition must be added. One iteration of the method takes 3 seconds on an IBM370/158 on this example.

A three dimensional nozzle is being tested: the result will be shown at the conference. 20 to 40 iterations are usually sufficient for the algorithm to converge. The results are in good agreement with the tabulated data. Simple and multi-bodies airfoils are also being tested. For them it is necessary to include the entropy condition; 80 iterations are usually more than sufficient for the convergence.

8. CONCLUSIONS

Thus this method seems very promising. It compares very well with the finite differences method available and it has the advantage of allowing complicated two and three dimensional geometries. This work illustrates the fact that optimal control theory is a powerful tool with unexpected applications sometimes.

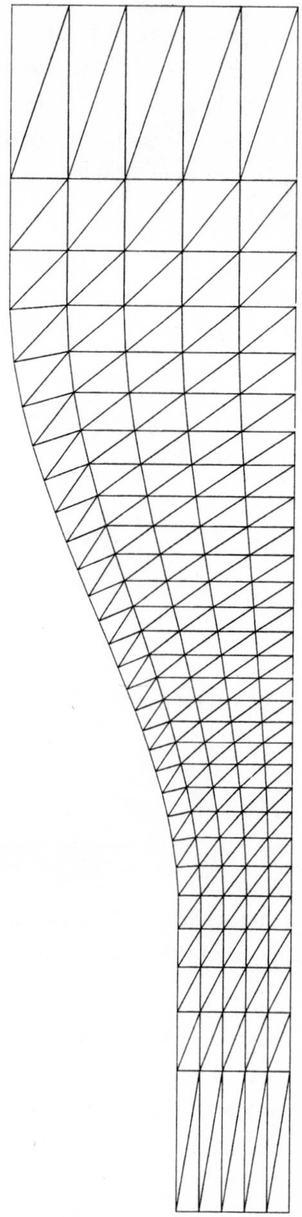

Figure 1: Discretization of the Nozzle

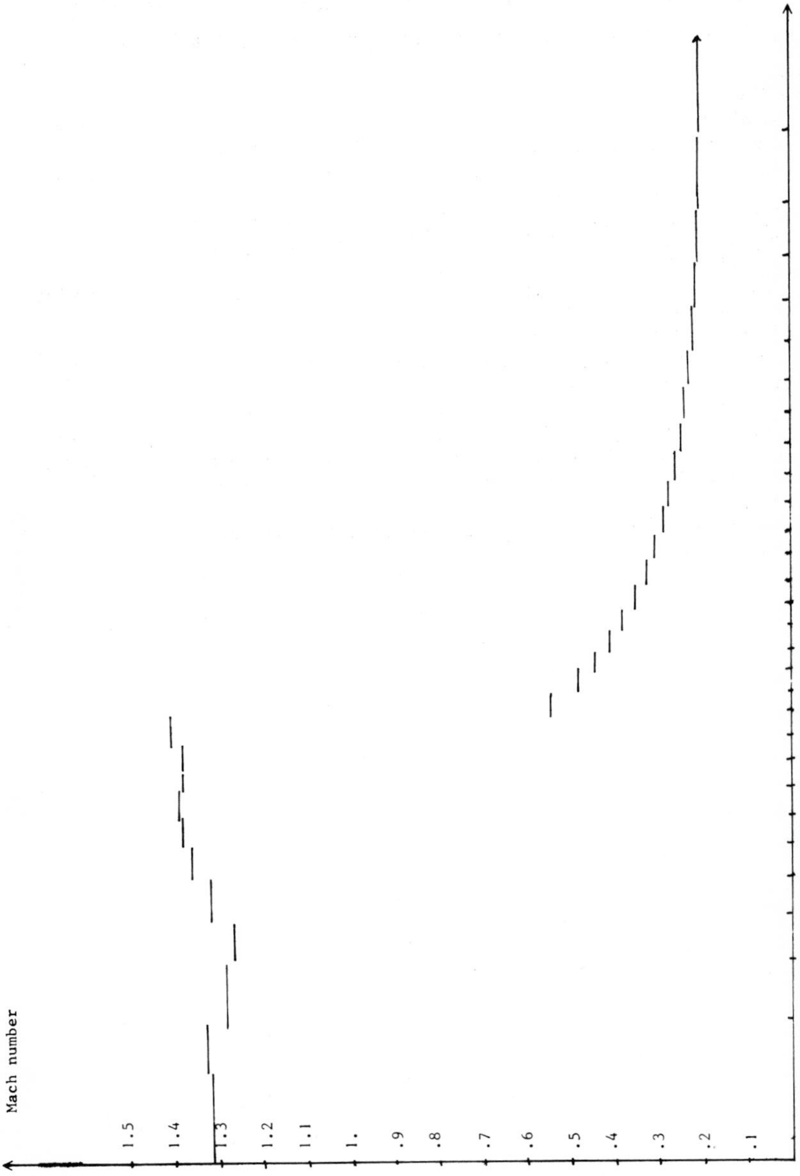

Figure 2: Axial Velocity on the Axis of the Nozzle

ACKNOWLEDGMENT

We wish to thank M. Periaux, Perrier and Poirier for allowing us to use their data files and computer, and for their valuable comments.

REFERENCES

1. Garabedian, P. R., Korn, D. G. - Analysis of transonic airfoils. Com. Pure Appl. Math., Vol. 24, pp. 841-851 (1971).
2. Gelder, D. - Solution of the compressible flow equation. Int. J. on Num. Meth. in Eng., Vol. 3, pp. 35-43 (1971).
3. Glowinski, R., Periaux, J., Pironneau, O. - Transonic flow computation by the finite element method via optimal control. Congrès ICCAD Porto Fino, June 1976.
4. Glowinski, R. and Pironneau, O. - Calcul d'écoulement transsonique par des méthodes d'éléments finis et de contrôle optimal. Proc. Conf. IRIA, December 1975.
5. Jameson, A. - Iterative solution of transonic flows. Conf. Pure and Applied Math. (1974).
6. Norries, D. H. and G. de Vries - The Finite Element Method. Academic Press, New York (1973)
7. Periaux, J. - Three dimensional analysis of compressible potential flows with the finite element method. Int. J. for Num. Methods in Eng., Vol. 9 (1975).
8. Polak, E. - Computational methods in optimization. Academic Press (1971).

A 7
B 8
C 9
D 0
E 1
F 2
G 3
H 4
I 5
J 6